# GOVERNANCE OF THE SMART MOBILITY TRANSITION

# GOVERNANCE OF THE SMART MOBILITY TRANSITION

EDITED BY

**GREG MARSDEN**
*University of Leeds, Leeds, UK*

**LOUISE REARDON**
*University of Birmingham, Birmingham, UK*

emerald PUBLISHING

United Kingdom – North America – Japan
India – Malaysia – China

Emerald Publishing Limited
Howard House, Wagon Lane, Bingley BD16 1WA, UK

First edition 2018



**British Library Cataloguing in Publication Data**
A catalogue record for this book is available from the British Library

ISBN: 978-1-78754-320-1 (Print)
ISBN: 978-1-78754-317-1 (Online)
ISBN: 978-1-78754-319-5 (Epub)

Printed and bound by CPI Group (UK) Ltd, Croydon, CR0 4YY

ISOQAR REGISTERED

ISOQAR certified Management System, awarded to Emerald for adherence to Environmental standard ISO 14001:2004.

Certificate Number 1985
ISO 14001

INVESTOR IN PEOPLE

# CONTENTS

# ABOUT THE EDITORS

**Greg Marsden** is Professor of Transport Governance at the Institute for Transport Studies at the University of Leeds, UK. He has researched the design and implementation of new policies for over 15 years, with a particular focus on climate and energy policy. His current work examines how the smart mobility transition will change how we think about what, how and who governs the mobility system. He is currently Chair of the Commission on Travel Demand, looking at how to deal with alternative demand futures. He is the Secretary General of the World Conference on Transport Research Society (WCTRS) and Co-Chair of the Special Interest Group on Governance and Decision-Making Processes. He has served as an advisor to the House of Commons Transport Select Committee and regularly advises national and international governments.

**Louise Reardon** is Lecturer at the Institute of Local Government Studies (INLOGOV), University of Birmingham, UK. Her research explores the implications of governance arrangements and public policy processes on decision-making, with a particular interest in transport and wellbeing. Her research is at the forefront of understanding the political and policy interest in wellbeing, and she recently co-authored a book entitled *The Politics and Policy of Wellbeing: Understanding the Rise and Significance of a New Agenda*. As Co-Chair of the WCTRS Special Interest Group on Governance and Decision-Making Processes, she is keen to grow the community of scholars critically engaged in understanding and challenging the status quo of transport policymaking.

# ABOUT THE AUTHORS

**David Ashmore** is Affiliate Researcher at the University of Melbourne. He is currently working towards his doctorate which examines the symbolic aspects of transport choice across different cultures. His professional background is in transport regulation and procurement; he has worked for consulting firms, universities and the civil service.

**Carey Curtis** is Professor of City Planning and Transport at Curtin University, Director of Urbanet research network and Guest Professor at the University of Gothenburg. Her research interests include city form and structure, transit-oriented development, personal travel behaviour, accessibility planning, institutional barriers to sustainable transport, governance and transport policy.

**Diane E. Davis** is Charles Dyer Norton Professor of Regional Planning and Urbanism, and Chair of the Department of Urban Planning and Design at the Harvard Graduate School of Design. Her published works examine the relations between urbanization and national development, the politics of urban policy and urban governance. Her current research focuses on the future of cities in an era of rapid technological innovation, climate change and new forms of sovereignty.

**Iain Docherty** is Professor of Public Policy and Governance at the University of Glasgow. He has held board-level appointments with client and provider sides of the transport industry in the United Kingdom and advised public agencies in the United States, Australia, Canada, the Netherlands and Sweden, and the OECD.

**Robyn Dowling** is Professor of Urbanism in the School of Architecture, Design and Planning at the University of Sydney. She is well known for her work on the cultural geographies of suburban homes and neighbourhoods. Her current research focuses on urban policy responses to technological disruptions, focusing on smart mobility and the implementation of smart city strategies.

**Edgar Salas Gironés** is a PhD Candidate in Innovation Governance at Eindhoven University of Technology. In his PhD project, he studies the emerging governance arrangements between private and public actors for the transition towards Smart Mobility in the Netherlands.

**Debbie Hopkins** is a Departmental Research Lecturer in the School of Geography and the Environment, University of Oxford. Her research responds to questions of behaviour change, socio-technical innovation, and low carbon transitions. She is the co-editor of *Low Carbon Mobility Transitions* (Goodfellow, 2016).

**Crystal Legacy** is Senior Lecturer in Urban Planning at the University of Melbourne. Her research examines questions of urban conflict and citizen engagement with a current focus on the role of the citizen in contested transport processes in Australian and Canadian cities. She is co-editor of *Instruments of Planning: Tensions and Challenge for More Equitable and Sustainable Cities* (Routledge, 2015).

**Dimitris Milakis** is Assistant Professor of Smart and Sustainable Transport Systems at Delft University of Technology. His research focuses on the influences of the built environment and (emerging) transport systems on human travel and location behaviour. He is also interested in human perceptions and preferences of travel and their integration into urban and transport planning.

**Miloš Mladenović** is an Assistant Professor at the Spatial Planning and Transportation Engineering Group, Aalto University. His current research interests include ethical assessment of emerging mobility technologies, socially sustainable transport planning methods, asset management methods for intelligent transport systems and transport engineering education practices.

**Kate Pangbourne** is an EPSRC/LWEC-funded University Academic Fellow at the Institute for Transport Studies, University of Leeds. She has an MA (Hons) in Philosophy with English Literature, an MSc in Sustainable Rural Development and a PhD in Geography (Environmental). With an interdisciplinary background encompassing environmental sustainability, transport geography, technology, social science and philosophy, she addresses changing behaviours, practices and governance.

**Jan Scheurer** is Senior Research Fellow at Curtin University, Perth, and an Honorary Associate at RMIT University, Melbourne/Barcelona. He has 20 years of academic experience in the fields of transport and accessibility planning, urban design and sustainability studies.

**Tim Schwanen** is Associate Professor of Transport Studies and Director of the Transport Studies Unit at the University of Oxford. He holds a PhD in Geography from Utrecht University, the Netherlands and has published extensively on various dimensions of everyday mobility, including the role of new technologies and the dynamics and prospects of low carbon transitions in urban mobility.

**Dominic Stead** is Associate Professor of Urban and Regional Development at Delft University of Technology and Honorary Research Fellow at University College London. He has held positions at the University of Queensland, Friedrich-Alexander University Erlangen-Nürnberg, HafenCity Universität Hamburg, University College London and the University of the West of England.

**John Stone** is Senior Lecturer in Transport Planning at the University of Melbourne. His research explores the political and institutional context for variation in international transport planning practice, with a focus on cities in Australia, Canada and German-speaking Europe. He has also worked in local government and as a community advocate for sustainable transport.

**Darja Vrščaj** graduated from Maastricht University in 2014, having completed a research MSc in Science and Technology Studies. In the past, she worked at the European Parliament, STOA Unit and the OECD, STI Directorate. Currently, she is pursuing a PhD at the Technical University of Eindhoven. Her research is co-funded by the Ministry of I&M.

# ACKNOWLEDGEMENTS

We would like to thank the International Transport Forum and in particular Jose Viegas, Stephen Perkins, Claire Millar and Magdalena Olczak for their support in organizing the research day and side event to the 2017 International Transport Forum Summit in Leipzig from which this collaboration grew.

# CHAPTER 1

# INTRODUCTION

Greg Marsden and Louise Reardon

ABSTRACT

*Despite the massive social benefits that the car has brought, it has become evident that the current mobility system is undermining the benefits it creates with substantial air quality problems, inactive lifestyles, deaths and injuries from accidents and major contributions to the global climate change challenge. The introduction of smart mobility innovations, in promising to challenge the existing regime of automobility may be a major policy opportunity, and also provide a source of new economic opportunity. However, it is far from clear that these opportunities will be recognized or, even where they are, realized due to the complexities of steering any transition in the mobility system.*

*This book sets out how we should understand the challenge of governing the smart mobility transition and, in this introductory chapter we set out the key arguments and contributions of each part of the book for addressing these challenges. The first section of the book focuses on how the role of the government is challenged by the growing network of actors and the new resource interdependencies that emerge from smart mobility. How these challenges come to be recognized and resolved is itself a critical part of the governance process as explored in the second section. The third section examines the changing context of governance and the capacity of the state to act to steer the transition. This allows us to identify, in our final concluding section, a set of critical topics for those researching and implementing the smart mobility revolution.*

**Keywords:** Governance; smart mobility; institutions; meta-governance

## WHAT IS SMART MOBILITY?

This book is about the governance of smart mobility. However, it is necessary to ask what smart mobility is before we can describe why it might be different or interesting to think about the governance of it. Is it actually anything new? Transport systems are always evolving and each generation can be seen to be imbued with more technology and therefore in some ways 'smarter' than what went before. For example, the introduction of the first traffic signal outside the Houses of Parliament in England in 1868 was the start of a new wave of rules, regulations and technologies which shape how traffic is managed in cities today. Smart mobility is therefore perhaps more a label with currency than anything specific. However, the label has prevalence perhaps because we are in a period where a series of innovations are being brought forward which separately and (potentially) in unison are promising to significantly change how people move around. Some would argue that they have already done this. To date, smart mobility innovations include:

- The ability and need to electrify the vehicle fleet using battery power, plug-in hybrid and/or other new technologies (Dijk, Orsato, & Kemp, 2013);
- Almost ubiquitous mobile Internet, mapping technologies and related 'apps' available in portable smartphone devices allowing two-way flows of data and information, integrating mobility and non-mobility options (see Toole et al., 2015);
- Increasingly autonomous vehicles that initially support but may subsequently remove the driver from even a supervisory role in the vehicle,

allowing occupants to do other tasks on the move (see Fagnant & Kockelman, 2015);

- A shift towards models of 'usership' rather than 'ownership' of vehicles with an evolution from organized car-sharing schemes through real-time ride-hailing apps, to more integrated products referred to as 'Mobility as a Service (MaaS)' (Jittrapirom, Caiati, Feneri et al., 2017) and
- Increasingly intelligent infrastructure which is capable of not only interacting with users and vehicles in real-time to adapt its service but also potentially change user behaviour through price signals, information or other incentives (Alam, Ferreira, & Fonseca, 2016).

We also note that these innovations are occurring as part of a much broader transition of large aspects of society: for example, major changes to retail with the rapid growth of Internet shopping (Ginsberg & Uygur, 2017) and changes to healthcare technologies and provision (Bauchner, Berwick, & Fontanarosa, 2016). All of these will, in turn, impact on what we travel for and how and where we access goods and services.

While it is possible and often necessary to focus down on the specifics of a given technology, from the perspective of this book we are interested in how these technologies are introduced and what changes as a result. To understand this, it is necessary to think about the technologies as part of a broader socio-technical system of mobility. Rotmans, Kemp, and van Asselt (2001, p. 16) identify such systems as having a 'set of connected changes, which reinforce each other but take place in several different areas, such as technology, the economy, institutions, behaviour, culture, ecology and belief systems'. In the last 50 years of the previous century, most clearly in developed countries, the mobility system has arguably become a system dominated by the car, creating what Urry (2004) describes as a self-reinforcing system of automobility.

Despite the massive social benefits that the car has brought, it has become evident that the current system is undermining the benefits it creates with substantial air quality problems, inactive lifestyles, deaths and injuries from accidents and major contributions to the global climate change challenge. For these reasons, transitions and mobility scholars have argued that the current system needs to be transformed towards a smarter and more sustainable one (Geels, Kemp, Dudley, & Lyons, 2012). The introduction of smart mobility innovations that can challenge the existing regime of automobility is therefore both a source of new economic opportunity (for some) and a major policy opportunity (if recognized as such).

At this early stage of smart innovation there are signs of positive changes, such as a shift away from car ownership towards more active modes, and greater use of public transport and ride-sharing services, particularly in the centres of major cities (e.g., Le Vine & Polak, 2017; Rabbitt & Ghosh, 2016). There are also impressive policy visions of transport systems where, decades from now, the completed transition to smart mobility has enabled circumstances in which there is no congestion on our streets and the ability to hand back urban spaces for parkland and social interaction is plentiful (ITF, 2017). However other, more challenging, outcomes are also possible. Increased automation could, for example, promote a much more individualized and personalized mobility system where there is significantly more traffic (Wadud, Mackenzie, & Leiby, 2016) and where such innovations further cement the longevity of automobility (Schwanen, 2016). For the most positive and optimistic visions to come about would require multiple innovations to succeed and a level of collaboration and cooperation which has hitherto not been a feature of transport markets which have tended instead to produce bespoke and poorly integrated transport systems. The behaviour of some of the early innovators in the system also suggests the potential to produce negative externalities, possibly even undermining rather than enhancing sustainable transport aims (Kitchin, 2015).

While there is inevitably uncertainty about the scale, pace and nature of the development of smart mobility, we start from the position that some and perhaps many of these innovations will prevail. Others may remain as niche innovations or disappear altogether. Others will surely emerge. The move to a system characterized by more 'smart mobility' is inevitable. What is less clear is how it will happen and how differently it will happen in different places. This we suggest is contingent on how state and non-state actors interact to shape the future transition or, in short, how smart mobility is governed. We now turn our attention to what we mean by governance.

## GOVERNANCE

Fundamental to the concept of governance is its analytical separation from the notion of government. As Stoker (1998, p. 17) notes, government refers to 'the formal institutions of the state and their monopoly of legitimate coercive power' and in particular 'the formal and institutional processes which operate at the level of the nation state to maintain public order and facilitate collective action'. In contrast, 'the essence of governance is its focus on governing mechanisms which do not rest on recourse to the authority and sanctions of government' (Stoker, 1998, p. 17). In turn, governance 'evokes a world in which state power is dispersed among a vast array of spatially and functionally distinct networks composed of all kinds of public, voluntary, and private organizations with which the centre now interacts' (Rhodes, 2011, p. 34).

Rhodes (1996, 2007) identifies four underpinning characteristics of governance. The first is the growing interdependence of organizations. The delegation of state authority to autonomous agencies, and the outsourcing of delivery of public services to private companies, for example, has meant the boundaries between state, private and voluntary sector organizations have 'shifted' and grown 'opaque' (Rhodes, 1996, p. 660). The second characteristic is the sustained interaction of members of networks due to their dependence on one another for resources (such as finance, information and legal authority), which mean they must negotiate shared goals. The third characteristic is 'game-like interactions' that are rooted in trust and regulated by 'rules of the game' that are negotiated through shared purposes. The fourth characteristic is 'a significant degree or autonomy from the state'. Rhodes (1996) argues that because networks are 'self-organizing' they are not accountable to the state, and therefore the role of the state is to 'directly and imperfectly steer networks'. This notion of 'steering' lays in contrast to 'rowing' in which the state provides and delivers policies and services itself (Osborne & Gaebler, 1992).

However, these characteristics of governance do not suggest in and of themselves that the outputs of governance should, or need to be, different from those of government. As Stoker (1998, p. 17) argues, 'governance is ultimately concerned with creating the conditions for ordered rule and collective action', and therefore the difference between government and governance is the process by which the outputs are achieved (Rosenau, 1992; Stoker, 1998, p. 17). As a consequence, many definitions of governance stress the provision of rules, laws and guidelines to determine action. For example, Kjaer (2004, p. 10) sees governance as 'the setting of rules, the application of rules, and the enforcement of rules', while Donahue (2002, p. 1) understands governance as 'the rules and institutions for the authoritative organization of collective life'. The extent to which government (or 'the state') has the power to set, apply and enforce these rules and how it does so is contested. This process of the 'governance of governance' is understood as 'meta-governance'. The literature points to the importance of framing, story-telling, support and facilitation of networks as important tools for meta-governance over and above more traditional hierarchical, top-down mechanisms (Sørensen, 2006; Torfing, Peters, Pierre, & Sørensen, 2012).

## SMART MOBILITY AND GOVERNANCE

This book was assembled from papers presented at a meeting of the World Conference on Transport Research Society Special Interest Group on Governance and Decision-Making, hosted as part of the International Transport Forum Summit on Governance in May 2017. We introduce the structure of the book and the contents of each chapter by drawing together the brief introductory literature reviews on smart mobility and governance and presenting some key themes.

The first section of the book, *Navigating the Role of the State*, explores the extent to which smart mobility might change and challenge the position of the state within the transport system. It is clear that smart mobility is bringing a new set of actors to the transport arena. These actors include global technology companies such as Google and Apple; (transport) service aggregators such as Uber and Lyft; and firms specializing in artificial intelligence, automation and robotics who are working with incumbent providers to change the nature of existing goods and services (for example, BMW's partnership with Mobileye). This expanded and increasingly complicated multi-level network of actors has the potential to change the nature of the state's role as a provider of services ('rowing' within the transport system), but also challenges its existing relationships and position in transport governance networks more broadly. For example, in 2004 the UK Government established a pioneering journey planning portal for the United Kingdom (Transport Direct), providing the information necessary to enable multi-modal journey planning. The portal closed in 2014 with the Government reporting that 'equivalent travel information services are now widely available online from several other sources' (DfT, 2014), marking a movement of the state away from information provision.

The business models of smart mobility companies are predicated on challenging incumbent providers in order to give consumers something new and 'better', which is why smart mobility innovations are often referred to as 'disruptive'. Subsequently, we are seeing well-established state–provider relationships being resisted, redefined and sometimes renegotiated (Dudley, Schwanen, & Baniste, 2017). For example, Uber has popular and successful operations in over 675 cities around the globe, but yet has faced wide-spread protests from incumbent operators (e.g., in London and Paris), is banned from some countries (e.g., Italy and Bulgaria) and has withdrawn operations or had its license revoked in other cities where they have refused to comply with regulations which apply to incumbent taxi and private hire companies (e.g., Copenhagen and Austin, Texas). This section of the book, therefore, begs questions of how the state should behave in these circumstances, reflects on whether there are areas where the state should change its role (such as in the Transport Direct example), and how and whether the state should justify interventions to limit the behaviour of firms such as Uber when so many people sign up and use them.

The first chapter in this section, by Iain Docherty, sets out the wider arguments for the key and unique role the state must play in the mobility system and asks to what extent these arguments are diminished as a result of new 'smart' innovations, or simply in need of revision and recalibration. Drawing reference to the wholly optimistic visions of the potential of the automobile from the 1950s, Docherty suggests that while smart mobility innovations might promise to deliver significant societal benefits for all, there also exist pathways which take us further away from sustainability. It is also necessary, he argues, to understand that just as existing private-sector transport operators provide uneven services both geographically and temporally due to their focus on profitable markets, so too will new mobility providers. Docherty argues therefore that the state's task is to develop and adapt in ways which continue to allow it to both set the over-arching direction of policy and have a sufficient hand on the tiller to steer towards positive social outcomes, part of which will include continuing to play an important role as guarantor of a certain level of socially necessary or desirable services.

The following chapter by Kate Pangbourne, Dominic Stead, Milos Mladenovic and Dimitris Milakis drills down further into these issues through the case of MaaS. The authors introduce the importance of framing as an aspect of state involvement. While there are in fact many different forms of MaaS, it is commonly assumed to be an innovation which reduces car ownership and unsustainable travel. However, as the authors point out, the business models are predicated on making travel cheaper and easier and on selling more, not less, mobility. Before jumping on MaaS as a solution therefore, they suggest much greater clarity is given to the ways in which MaaS might be supported or regulated to ensure the systems contribute sufficiently in the long run to the public good and provide adequate consumer protection.

The second section of the book asks, '*Whose Voices Are in the Smart Mobility Debate?*' As Rhodes argues, continuing interaction between members of the governance network is one of the ways in which shared goals can be reached. However, there exist asymmetries of power within networks; not all members are equally reliant on the resources of others to achieve their goals, and therefore not all voices receive equal attention. The question of who gets to participate is, therefore, critical to understanding how agendas surrounding smart mobility, its implementation and outcomes will emerge. It is important here to state that, while each of the chapters in this section critically reflects on different aspects of the changing voices in the mobility debate, the direction of travel should not necessarily be inferred as negative. After all, the outcomes of the current [auto] mobility system do not work for everyone and privilege some communities of actors while marginalizing others (Bickerstaff & Walker, 2005). Smart mobility may provide opportunities to resolve, or mitigate, some of these issues.

Chapter 4, by Robyn Dowling, exposes the importance of issue framing for determining how networks are accessed and shaped. Through a case study approach, she demonstrates that the legally defined or socially accepted categories of what transportation services are (e.g., what delineates 'public transport') defines the extent to which innovations fit or disrupt existing governance systems. The state can easily categorize some innovations, such car sharing, and in turn adapt existing tools and practices to allow these new uses to emerge and flourish. Other innovations, such as driverless cars, services such as Uber, and new products such as Segways, are harder to categorize as they have no precedent. It is therefore a struggle for new innovators to make their way into the system, and so they adopt a strategy of framing themselves in a way that may sometimes obscure their intended purpose or potential. Who has the power to set, shape or resist these categorizations is a crucial and understudied part of the smart mobility governance story.

Chapter 5, by Debbie Hopkins and Tim Schwanen, challenges the premise that smart mobility transitions can be developed effectively through existing, market-orientated governance approaches, which exemplify a 'post-political' narrative of transport provision in which the market will provide the necessary solutions. Drawing on the case of vehicle automation, they point to the very uncertain nature of technological developments and societal acceptance of such developments as requiring a more reflexive and participatory style of governance, as recognized in the Transitions Management literature. They argue that successful implementation will need to occur through experimentation and social participation as well as through user, provider and governmental learning within and across networks, and therefore requires a governance environment in which the state and those that experience smart innovations continue to actively shape the systems they need.

Chapter 6, by Edgar Salas Gironés and Darja Vrščaj, takes a case study approach to ask *Who Benefits from Smart Mobility Policies?* Analysing two schemes designed to reduce congestion through the acceleration of e-bike adoption, their chapter looks at the important question of how new innovations are framed and the effect of these framings on different groups. Drawing on frameworks from the Social Construction of Policy Design and Science and Technology Studies, they identify that the benefits of the scheme are aligned to those who currently commute to work by car, while students and those who already take non-car-based modes to work are marginalized. These findings are representative of a more general challenge to accelerating the adoption of many new innovations, which is that they tend to focus on the wealthier early adopters. This does not have to be the case for smart mobility, but again serves as a reminder of the conscious and unconscious biases in planning where problems are solved for 'people like us' (Ralph & Delbosc, 2017), and which therefore marginalizes other groups with different and sometimes greater need.

The third section of the book focuses on *State Capacity*, or the ability of the state to exercise its power within a network of actors in order to 'set the rules of the game' and steer policies towards chosen outcomes. Here, we draw on Pollitt's (2008) reminder that time is an understudied facet of political science and implementation studies, and plays a role in determining the ebbs and flows of power and where it resides. For example, the smart mobility transition has developed during a period of significant fiscal re-adjustment following the global financial crisis of 2008. The fiscal re-adjustment has led to widespread and substantial reductions in government expenditure, for example, local governments in England have experienced a real terms cut in revenue of 36.3% between 2009–2010 and 2014–2015 (IFS, 2015, p. 1). As Gardner (2017, p. 158) notes, such pressures have led to 'market-driven approaches to co-ordination…growing in importance in comparison to state-driven models', for example, 'the exploration of outsourcing in relation to specific functions…'. However, while the financial crash may have increased the urgency with which such changes are being implemented, they are not new and are part of a longer process of trends towards New Pubic Management in the context of political projects which stress the efficiency that can be gleaned from a 'small-state' and ever-greater delegation of tasks to the private sector and other non-state providers (Bovaird & Loeffler, 2016; Hood, 1991). It is within this context then that we have to understand where within government there is the capacity to steer the mobility transition and what factors contribute to building rather than eroding that capacity.

Chapter 7, by Diane E. Davis, explores how two different smart mobility innovations came to be implemented in the way they were and what this has meant for the development or diminution of state capacity. In so doing, Davis points to the need to pay attention to styles of governance and to the role of policies as a tool of governance. In the case of the introduction of Uber in San Francisco, where there was no incumbent policy framework applicable to Uber, the local government initially sought to block its activities and was outflanked at a state level where the power of opposition was weaker and the economic influence of technology companies felt more strongly. The legacy is weaker local government control over mobility, and little development of capacity at the State level.

By contrast in Stockholm, congestion charging was driven by the state, albeit through a challenging political process. Here though, the policy was state-driven and brought in private-sector 'smart' expertise to improve the scheme's delivery and reputation of the city, which facilitated trust between the state, smart mobility operator and users, enabling further innovations to be implemented.

Chapter 8, by John Stone, David Ashmore, Jan Scheurer, Crystal Legacy and Carey Curtis, explores some of these themes through a different setting. The Australian case warns of the long-term impacts of neo-liberal, market-driven transport provision and a subsequent closing off of mindsets to the potential for delivering innovations through collective, state-led, effort. The on-going development of privately financed road infrastructure investment is symptomatic, they argue, of a 'hands-off' mentality that might limit the extent to which the state would wish to (or be seen to) influence the roll out of technologies such as autonomous vehicles. Through interviews with public sector officials, the authors demonstrate that there is a very low level of strategic thought and capacity going in to planning for a smart mobility transition despite an awareness that it matters to their existing portfolio of operations and practices. The authors describe an 'existential crisis of legitimacy and purpose' for the state which will be a major challenge to any notion of intentional steering of a transition.

Chapter 9, by Greg Marsden and Louise Reardon, explores the importance of the type of role the state takes in steering the smart transition through a scenario exercise conducted with experts from nine different countries. The scenario exercise was set up to understand which smart mobility innovations would 'flourish' or 'falter' in a two-dimensional scenario space. The dimensions chosen were a more directive and 'hands-on' governance at one end of the axis to a more laissez-faire 'hands-off' style at the other, and high social acceptance of innovation through to low social acceptance on the other axis. The experts were able to identify different cities or countries as being more aligned with some of these scenario spaces today. It was felt that only in the hands-on, social acceptance space would a fully integrated and multi-modal system of smart mobility emerge. In other spaces some innovations were seen to be attractive and likely to be deployed but, in general, on a more individualistic mobility model. The exercise points to a key argument in the book, which is that the role of government within a broadening network of governance matters significantly to the sort of smart mobility transition that will be delivered.

Chapter 10, by Louise Reardon and Greg Marsden, draws the key themes of the book together, focusing on why it is necessary and what it will take for the state to engage with, steer and build capacity as part of the expanding governance networks of the smart mobility world we are moving towards. It does so through further reflection on the eight contributory chapters. We argue that, although early days in the transition, it is necessary to start considering now what thinking and adaptation is necessary to get the best out of the technologies being developed. It is though still the start of that journey and we hope this book provides the start of many further critical debates to advance the study of the governance of smart mobility.

## REFERENCES

Alam, M., Ferreira, J., & Fonseca, J. (2016). *Intelligent transport systems*. New York, NY: Springer.

Bauchner, H., Berwick, D., & Fontanarosa, P. B. (2016). Innovations in health care delivery and the future of medicine. *JAMA*, *315*(1), 30–31.

Bickerstaff, K., & Walker, G. (2005). Shared visions, unholy alliances: Power, governance and deliberative processes in local transport planning. *Urban Studies*, *42*(12), 2123–2144.

Bouvaird, T., & Loeffler, E. (2016). *Public management and governance* (2nd ed.). Abingdon: Routledge.

DfT. (2014). Transport Direct website closes on 30 September 2014, Gov. UK. Retrieved from https://www.gov.uk/government/news/transport-direct-website-closes-on-30-september-2014. Accessed on 29 October 2017.

Dijk, M., Orsato, R. J., & Kemp, R. (2013). The emergence of an electric mobility trajectory. *Energy Policy*, *52*, 135–145.

Donahue, J. D. (2002). Market-based governance and the architecture of accountability. In J. D. Donahue & J. S. Nye Jr. (Eds.), *Market-based governance: Supply side, demand side, upside and downside.* Washington, DC: The Brookings Institute.

Dudley, G., Schwanen, T., & Banister, D. (2017). The rise of Uber and regulating the disruptive innovator. *The Political Quarterly*, *88*(3), 492–499.

Fagnant, D., & Kockelman, K. (2015). Preparing a nation for autonomous vehicles: Opportunities, barriers and policy recommendations. *Transportation Research Part A: Policy and Practice*, *77*, 167–181.

Gardner, A. (2017). Big change, little change? Punctuation, increments and multi-layer institutional change for English local authorities under austerity. *Local Government Studies*, *43*(2), 150–169.

Geels, F. W., Kemp, R., Dudley, G., & Lyons, G. (Eds.). (2012). *Automobility in transition? Socio-technical analysis of sustainable transport*. UK: Routledge.

Ginsberg, A., & Uygur, A. R. (2017). Changing technology in transportation: Automated vehicles in freight. University of Illinois. Retrieved from https://ntl.bts.gov/lib/62000/62200/62290/Automated-Vehicles-in-Freight-Revised-Draft-06272017.pdf

Hood, C. (1991). A public management for all seasons? *Public Administration*, *69*(1), 3–19.

IFS [Institute for Fiscal Studies]. (2015). Central cuts, local decision-making: Changes in local government spending and revenues in England, 2009–10 to 2014–15: IFS Briefing Note BN166. Retrieved from https://www.ifs.org.uk/uploads/publications/bns/BN166.pdf

International Transport Forum Corporate Partnership Board. (2017). Transition to shared mobility: How large cities can deliver inclusive transport services. OECD-ITF Corporate Partnership Board Policy Insights, June. Retrieved from https://www.itf-oecd.org/sites/default/files/docs/transition-shared-mobility.pdf

Jittrapirom, P., Caiati, V., Feneri, A.-M., Ebrahimigharehbaghi, S., Alonso González, M. J., & Narayan, J. (2017). Mobility as a service: A critical review of definitions, assessments of schemes, and key challenges. *Urban Planning*, 2(2), 13–25.

Kitchin, R. (2015). Making sense of smart cities: Addressing present shortcomings. *Cambridge Journal of Regions, Economy and Society*, 8(1), 131–136.

Kjaer, A. M. (2004). *Governance*. Cambridge: Polity.

Le Vine, S., & Polak, J. (2017). The impact of free-floating car sharing on car ownership: Early-stage findings from London. *Transport Policy*, in press. doi: https://doi.org/10.1016/j.tranpol.2017.02.004

Osborne, D., & Gaebler, T. (1992). *Reinventing government: How the entrepreneurial spirit is transforming the public sector*. London: Penguin Books.

Pollitt, C. (2008). *Time, policy, management: Governing with the past*. Oxford: Oxford University Press.

Rabbitt, N., & Ghosh, B. (2016). Economic and environmental impacts of organised car sharing services: A case study of Ireland. *Research in Transportation Economics*, 57, 3–12.

Ralph, K., & Delbosc, A. (2017). I'm multimodal, aren't you? How ego-centric anchoring biases experts' perceptions of travel patterns. *Transportation Research, Part A: Policy and Practice*, *100*, 283–293.

Rhodes, R. (1996). The new governance: Governing without government. *Political Studies*, 652–667.

Rhodes, R. (2007). Understanding governance: Ten years on. *Organization Studies*, *28*(8), 1243–1264.

Rhodes, R. (2011). Policy network analysis. In D. Levi-Faur (Eds.), *The Oxford handbook of governance* (pp. 425–447). Oxford: Oxford University Press.

Rosenau, J. (1992). Governance, order and change in world politics. In J. Rosenau & E. O. Czempiel (Eds.), *Governance without government: Order and change in world politics* (pp. 1–30). Cambridge: Cambridge University Press.

Rotmans, J., Kemp, R., & van Asselt, M. (2001). More evolution than revolution: Transition management in public policy. *Foresight*, *3*(1) 15–31.

Schwanen, T. (2016). Rethinking resilience as capacity to endure. *City*, *20*(1), 152–160.

Sørensen, E. (2006). Metagovernance: The changing role of politicians in processes of democratic governance. *The American Review of Public Administration*, *36*(1), 98–114.

Stoker, G. (1998). Governance as theory: Five propositions. *International Social Science Journal*, *50*(155), 17–28.

Toole, J., Colak, S., Sturt, B., Alexander, L., Evsukoff, A., & González, M. (2015). The path most traveled: Travel demand estimation using big data resources. *Transportation Research Part C: Emerging Technologies*, *58*, 162–177.

Torfing, J., Peters, B. G., Pierre, J., & Sørensen, E. (2012). *Interactive governance: Advancing the paradigm*. Oxford: Oxford University Press.

Urry, J. (2004). The 'system' of automobility. *Theory, Culture and Society*, *21*, 25–39.

Wadud, Z., MacKenzie, D., & Leiby, P. (2016). Help or hindrance? The travel, energy and carbon impacts of highly automated vehicles. *Transportation Research Part A: Policy and Practice*, *86*, 1–18.

# SECTION ONE

# NAVIGATING THE ROLE OF THE STATE

# CHAPTER 2

# NEW GOVERNANCE CHALLENGES IN THE ERA OF 'SMART' MOBILITY

Iain Docherty

ABSTRACT

*This chapter seeks to bring more clarity and urgency to the debate about the impacts of 'smart' mobility by highlighting the critical role of the state in managing the transition to a 'smart' transport future. This debate is currently dominated by producer-led imaginings of how technological innovation will solve apparent mobility problems, but has yet to address substantively more complex questions about how the impacts of these innovations will disrupt the mobility status quo, and how changing patterns of mobility in a 'smart' future will redefine the places in which we live. The chapter explores what impacts such innovations might have on the economic, environmental and social outcomes associated with the mobility system, and some key issues that systems of governance will have to address as 'smart' options become an increasingly important part of overall mobility provision.*

**Keywords:** Governance; transition; regulation; the state

## WE'VE SEEN THE FUTURE … AND IT WORKS

> *… if cars could drive themselves, there would be no need for most people to own them. A fleet of vehicles could operate as a personalized public-transportation system, picking people up and dropping them off independently, waiting at parking lots between calls. … Streets would clear, highways shrink, parking lots turn to parkland.*
>
> — Bilger (2013)

This quote, from Sergey Brin, President of Google's parent company Alphabet, demonstrates the scale of the utopian vision heralded by corporate proponents of the 'Smart Mobility' future set out in Chapter 1. If such a vision comes to fruition, advanced technology will transform the performance of the mobility system, with key innovations such as autonomous vehicles and the on-demand matching of the supply and demand of mobility services at the individual level opening up new social and economic opportunities and transforming road safety, which remains one of the most important public health challenges globally: for example, Fagnant and Kockelman (2015) estimate the potential societal benefit of each autonomous vehicle introduced as between US $2000 and US $4750 per annum.

However, as with any transition of this magnitude, the benefits will only be realized if the process of implementing smart mobility is carefully governed. Not only are the technologies involved new and disruptive, but some of the providers are new too (at least to the mobility domain). As a result, the power relations between the state and these market actors are already evidently different to those of the status quo. Nonetheless, at the same time as the state is facing the disruption of established 'rules of the game' for the regulation of mobility, some of the public policy challenges that governments have found difficult to address for decades, such as the exploitation of imperfect competition to create monopoly provision of services, are re-establishing themselves. The question facing the state is therefore not *whether* to react to the smart transition but *how* to react to it. In other words, what posture should governing institutions and networks adopt in the face of these new and not so new challenges so that perverse outcomes are eliminated, or at the very least substantially controlled?

This chapter, therefore, explores how the smart mobility transition might be governed such that its benefits are maximized and disbenefits minimized. If the state proves unable to effect its own transition to govern effectively in the smart era, then we run the risk of some quite substantial negative externalities becoming entrenched. Or, as a former New York traffic commissioner Janette Sadik-Khan noted,

> *without a regulatory framework, cities could see outcomes (from Smart Mobility) that run counter to goals of mobility, sustainability, accessibility and social equity. (Sadik-Khan & Solomonov, 2016, p. 285, brackets added)*

While this statement might seem rather obvious, consider that although almost every transport planning textbook opens with a version of the truism that 'transport (or mobility) is a derived demand', a worryingly large proportion of transport professionals and policymakers seem to forget this for their entire career in the sector, focusing instead on the delivery of 'wishlists' of transport *projects* that have more mobility as their core objective. Similar myopia is already evident in many attitudes to the smart transition. For example, Wockatz and Schartau (2015) collated a range of reasons why the state might be interested in intervening in the smart transition, many of which define support for smart mobility as a means to support an emerging technology sector, rather than necessarily something that might improve economic, environmental or social outcomes through better planning of mobility itself. The problem with this generally 'enabling' posture that Wockatz and Schartau identify as the usual approach of the state in their review is that it is at best a very 'hands-off' approach that is (deliberately) blind to the concept that there might be negative externalities and market failures resulting from this scale and pace of change. At worst, the state even succumbs to technology and producer interests, which seek to emphatically change the global narrative about what mobility is for, how it should be distributed across society and, by logical extension, how we should live our lives. This is often linked to a post-recession industrial strategy narrative as discussed by Hopkins and Schwanen in Chapter 5, but it also reflects longstanding concerns that the state has often been guilty of failing to allocate both capital and revenue resources efficiently in the transport domain, and has been slow to recognize the potential of innovation to improve service quality within this.

One longstanding critique of this highly laissez-faire approach to significant socio-technical innovation is Anthony Giddens' notion of the 'ensuring state' that configures itself and acts to ensure that democratically agreed policy goals are achieved:

> *The idea of the enabling state suggests that the role of the state is confined to stimulating others to action and then letting them get on with it. The ensuring state is an enabling state, but one that is expected or obligated to make sure such processes achieve certain defined outcomes. (Giddens, 2008, p. 9)*

In conceptualizing how the state should react to the 'enable or ensure' challenge, it is critical to understand that the business models of most 'smart' transport providers depend on the generation of *as much mobility as possible* so that their returns are maximized. Although rhetoric of the need for fewer vehicles, with empty freeways being turned into city parks abound in the imaginings of the smart future seen in Brin's and others' technology conference pitches, an autonomous vehicle is only making money if it is moving around carrying passengers and/or goods from place to place, and so will need to be in use as much as possible. Herein lies the great paradox of the smart future as currently postulated: while the transition is sold as being about reducing congestion and the number of individual vehicles in the system, the economic imperatives of the transformation are that there will be more mobility per head, and therefore *more traffic* on the roads in many real-world situations. For example, the International Transport Forum (2017) undertook a modelling exercise of Lisbon to envisage how mobility would change in a typical European city with full smart adoption and sharing. The somewhat heroic assumptions made in the model are extremely enlightening in terms of thinking through what would be required to ensure that a smart, autonomous mobility system actually achieved the kind of reduction in road traffic that it was sold as delivering. They were:

- Private cars are *entirely* replaced by different forms of ride or vehicle sharing with high levels of service across the day.
- There is only *one* overall system optimizer/manager that allocates the supply of vehicles to passenger demand.

- There are only *one or two* models of provision of shared mobility which *everyone* uses.
- *All users* accept whatever sharing of vehicles is imposed upon them by the system manager.

Even a cursory examination of these assumptions demonstrates how fragile they are in the reality of a democratic market society. Why should we expect (or indeed want there to be) only *one* mobility service provider? Why do we think that people used to travelling by themselves in their personal BMW will take kindly to the imposed sharing of a Hyundai? The answer, of course, is that we do not expect everybody will regard a shared smart mobility system as preferable to their existing choices. However, when the ITF's model was run to recognize a potential middle ground, that of the smart system achieving a 50% replacement of private cars in the city, overall road traffic *doubled*.

Even from early modelling work, it is already clear, therefore, that no amount of smart technology (or big data) will overcome the need for good policy, planning and governance of the smart transition. Although this is another statement that might appear axiomatic, it is important to consider the extent to which much transport policy over recent decades has been dominated by the need to address the negative externalities of the motor vehicle. Thus, although Colin Buchanan famously wrote in 1963 that 'the motor vehicle is a remarkable invention, so desirable that it has wound itself inextricably into a large part of our affairs. There cannot be any going back on it ...', we have nonetheless expended great effort and energy trying to go back on it at least in part, as the negative externalities of congestion, pollution, crashes, blight produced by roads and social exclusion caused by a lack of mobility opportunities in otherwise car-dominated places have become real and enduring policy concerns due to their impact on the economy, environment and quality of life.

Buchanan is often (deliberately) misunderstood as having *promoted* the ubiquitous use of the car in *Traffic in Towns* (Ministry of Transport, 1963), but his landmark publication did, in fact, offer a really quite sophisticated review of what the negative externalities of mass car use might be in different kinds of places, and also how failing to address them would cause significant harm. If the scale of the potential disruption to the mobility status quo presented in any of the promoted visions of smart mobility is even roughly correct, then the smart transition will be one as profound as that of the car, not just in terms of the technologies used to provide mobility, but also in terms of the economic, environmental and social changes that will occur as a result of these technological shifts and the consequent emergence of a new usership/'Mobility as a Service'-type model. Indeed, those corporate evangelists for a smart mobility future such as Brin who are currently heavily promoting the testing of autonomous vehicles and other smart technologies could easily be imagined as saying 'smart mobility is a remarkable invention, so desirable that it has wound itself inextricably into a large part of our affairs. There cannot be any going back on it'. Therefore, in 2017 as in 1963, it is just as incumbent upon researchers and practitioners grappling with the potential implications of the smart mobility transition to think through what the negative externalities of the transition might be, and how these could be managed and ameliorated, before we are locked into a path that is difficult to turn back on.

## WE'VE SEEN THE FUTURE ... AND IT MIGHT NOT WORK AFTER ALL

The academic literature is replete with policy analyses of how the promised benefits of mass motoring have not always played out as positively in practice as its proponents would have had us believe (see, for example, Docherty & Shaw (2011) and Schöller-Schwedes (2010) on the United Kingdom and Germany, respectively). A key subset of this work is the literature on transport governance, which provides myriad examples of how many governments have often struggled to effect intellectually simple changes in the balance of transport provision, such as reducing congestion through public transport priority, in practice in the face of political and provider opposition (see Gray, Docherty, & Laing, 2017; Marsden & Stead, 2011). Given that the governance of transport has not always been effective even with a relatively stable set of modes and infrastructures, it is instructive to set out how the task of transport governance might need to evolve in response to the technological and other changes characterizing the 'smart' era. Three key examples that highlight this need are:

1. Systems and networks of transport governance that are already weak are being further disrupted – or deliberately attacked – by several new and powerful players such as the global computing giants and agile service providers whose business models depend on the creative destruction of existing regulatory frameworks.

2. As has already notably occurred in the property rental market, consumers can become providers too in the 'sharing economy', which means they have a different (more demanding) set of expectations about how mobility is organized and taxed, and also that the inequalities between people with different levels of access to mobility can polarize further as owning a car becomes an easy way of securing rental income.

3. Governments and other public sector actors that are used to exerting regulatory control through (exclusive) ownership and management of data are having this control eroded as the data on which smart mobility depends is crowd sourced and/or generated by automated sensors.

Given the scale of these governance and regulatory challenges, there is no time to be lost in considering what will be required in order to recast our systems of governance to ensure the adoption of smart mobility is a positive transformation. Critical to this is the need to foster a wide-ranging democratic debate about what economic, environmental and social *outcomes* we wish to see as the objectives of the smart transition, and how we will approach the inevitable trade-offs between the specific costs and benefits that arise. In Chapter 5 Hopkins and Schwanen elaborate further on this with a call for more reflexive forms of governance. Perhaps most important of all in this debate is the question of whether we are content for the de facto purpose of the smart transition to be the maximization of mobility such that any and all demands for travel are met. Reducing the need to travel is a long-standing objective of transport governance interventions, especially with regards to congestion and air pollution. But in the rush to demonstrate – or at least believe – that smart transport will in future somehow avoid these problems, given its claimed potential for ultra-high efficiency and use of 'clean' power, we risk generating a new set of unintended consequences and potential policy failures in which the state will have a direct interest, not least because it will be required to 'solve' them.

Some examples of these unintended consequences are already apparent, such as the differential access to existing on-demand transport services by race and social class, and the need to give over prime public space so that autonomous vehicles can stop to pick up and set down passengers in sufficient proximity that their 'door-to-door' service promise is realized. But there will be many other externalities of the kind of hyper mobile culture that smart transport proponents foresee. Apple or Google might envisage a world in which you swap your train commute and night out at the city centre restaurant and cinema after work for a 39-minute autonomous taxi ride to your home 100 miles away, followed 45 minutes later by another automated Deliveroo vehicle bringing your favourite chef's meal to your door from the next city, which you sit down to eat while streaming the latest blockbuster on your giant TV. All very technologically impressive this may be, but the implications of the mass adoption of this kind of lifestyle in terms of vehicle kilometres, the potential for extreme residential sprawl and a new wave of the hollowing out of cities are truly enormous.

It is therefore essential to approach the smart mobility transition critically in order to think through what the changes it brings might look like in practice, so that potentially damaging path dependencies can be avoided, but also so that its undoubted potential benefits such as radically improved safety can accrue as quickly as possible. To do this will require a change in narrative from governments away from the current mantra of supporting and encouraging the implementation of smart transportation technologies in pursuit of the holy grail of more and more 'innovation', to a more balanced position that recognizes there are trade-offs in the socio-technological transition of smart mobility as in any other.

## FORK IN THE ROAD

If, as seems at least possible, the smart mobility transition has as much disruptive potential as the introduction of the automobile itself, then highly effective governance processes, as democratic arbiters both for debate about the goals of the transition and for the regulation of smart mobility provision to ensure whatever ground-rules for its operation are agreed, are paramount. Until now, producer interests have been successful in framing the smart transport future as something that we should want, or indeed something that we *need*, focusing on a small number of attractive (real and important) benefits as the motivation for the transition, and justification for state support of those companies leading it. However, such is the scale of the multi-generational choices in areas such as land use and settlement structure that we face in deciding how to implement smart mobility that we are at a critical juncture for transport governance:

> *(Critical junctures are) relatively short periods of time during which there is a substantially heightened probability that agents' choices will affect the outcome of interest. (Capoccia & Kelemen, 2007, p. 348)*

The public debate about smart mobility, and government's role in it, urgently needs to move to the stage where we openly discuss how we want our societies to evolve as a result of the introduction of these new services, as such it is the anticipated pace of technological change that seemingly minor incremental decisions can lock us in to particular trajectories for years to come. Providing transparency about the tensions between short-term and long-term outcomes resulting from particular policy choices is one of the most important roles of the state, given competitive pressures for quick financial returns, and then market dominance. This means that governments will need to articulate the impacts of smart mobility on complex issues such as land use planning and the social inclusion/polarization that can result from shifting patterns of accessibility resulting from transport innovation; impacts that are measured over decades. Governments will also need to radically change the way we think about the pricing and taxation of transport, since many traditional revenue raising frameworks based on the existence of a fleet of individually owned road vehicles fuelled by hydrocarbons will look hopelessly out of date even part way through the smart transition: for example, Lindberg and Fridstrøm (2015) estimated that one-third of existing transport-derived taxation revenue will disappear within 15 years simply from the switch from fossil fuels to electric traction.

Addressing the taxation issue will, further, inevitably open up a range of difficult conversations about the relationship between the state and the global corporations that are seeking to control mobility provision by acting as 'mobility aggregators'; i.e., the hubs where the big data that matches demand for travel with the supply of vehicles is managed. Are we content for a small number of the largest corporations to assume market dominance and control of public data in this way? Given these corporations' track record (see Dowling, 2014), is there any guarantee that the state will be able to tax mobility effectively in future when faced with a small number of global corporations acting in an oligopolistic manner? At the service interface level, will we create a new 'mobility divide' based on the existing 'digital divide' between rich and poor, and will private-sector smart mobility providers care?

All of these questions reflect the fact that mobility is facing up to the same philosophical debate about the future as is already apparent in other domains of public policy: put simply, how 'public' or 'open' do we want the future to be (Paget-Seekins & Tironi, 2016)? Do we want a closed – or to use the computing giants' preferred term, 'curated' – system in which those that can afford it choose from a very small number of (highly attractive and efficient) door-to-door on-demand services based on individual vehicles? Or should we seek to operationalize smart transport (and especially autonomous vehicles) as part of a much more open wider system with appropriate regulation that retains fixed public transport systems as the planned backbone of the city to avoid further sprawl, celebrates the flexibility of multimodalism, seeks to protect the conditions in which socially inclusive access to mobility is guaranteed and privileges the active modes above motorized transport to protect public spaces and promote public health?

As will be apparent, this chapter argues that the latter solution is preferable to the kind of unregulated, hyper mobile future which is implicit in many current producer-led imaginings of smart transport. This will not be easy, as governance of a transition of such complexity will require skilled alignment of many different interests in a highly dynamic technological and policy environment. Following Giddens' idea of the 'ensuring state', government will first have to propose and implement clear regulatory frameworks for critical underpinnings of the smart transition, perhaps most immediately data sharing, so that impenetrable barriers to new participants in the smart transport future do not emerge. This key standards and co-ordination role will need to be agreed alongside a transparent approach to the taxation of smart mobility, as changes to the revenue structure will inevitably need to be made incrementally, and will shape patterns of provision in price- and coverage terms, and therefore the equality of access to new services. All of this will require strong political accountability, meaningful public participation and the recognition by providers that they are critical stakeholders in a wider complex system. The importance of the state taking actions now in building capacity through managing the early stages of the transition is discussed further by Diane E. Davis in Chapter 7.

## CONCLUSIONS

The transition to a 'Smart Mobility' system is already well underway, and will take some time to mature. In much the same way as the car has come to dominate mobility in most places in the developed world, we can reasonably expect the mobility system to become 'smart' in most places eventually. However, just as the impacts of the car, its place in the mobility mix and the distribution of its costs and benefits vary quite considerably between different places – so we can expect the form of 'smart mobility', and the distribution of its costs and benefits, to be unequally shared.

Thus even if the smart transition happens as we're told it will by its proponents – the latest *fait accompli* for technological progress that we will find irresistible – its impacts will be very uneven in time and space. Despite the rhetoric, there will be winners *and* losers, and this will become apparent long before the transition has reached its promised end game of delivering ubiquitous clean green transport on demand (see also Chapter 6). The first stage of effective governance of Smart Mobility is, therefore, that the messiness of the transition in itself requires dealing with, so that inequalities in the spread of its benefits are not locked in, something that is rarely, if ever, acknowledged (although see Meadowcroft (2009) for an excellent discussion of these issues in the energy policy domain). The terrain for this challenge could hardly be more difficult to negotiate, as it is occurring at the time when some very significant cleavages are becoming apparent in many advanced democracies, which bring with them profound implications for how the state balances competing policy demands (see Marsden & McDonald, 2017). Prime amongst these are 'open versus closed' views on migration and economic development; demographic cliff edges where highly unequal distribution of wealth generates a profound rethinking of or crisis in the relationships between communities, places and generations; and perhaps most importantly the 'culture wars' between entrenched and highly contrasting world views and political preferences that make it difficult to have a nuanced discussion about what is legitimate for the state to do, and what it should recuse itself from doing.

The key question raised by this chapter, therefore, concerns how we ensure that we create and enhance public value – or quality of life – though the course of the smart mobility transition. Smart providers want to maximize our consumption of mobility to maximize their own returns, and at the same time externalize as many of the costs as possible onto others. We should not be surprised by this, as this is how market capitalism works, and why the state exercises its power through regulation to ameliorate these externalities and ensure as far as possible that they are distributed fairly, while allowing market innovation and competition to create new wealth. Yet most current analyses of the implications of the smart transition are curiously uncritical, even naïve, in their understandings of the motivations of smart mobility proponents. Although there are undeniably real potential gains to be had, especially in respect of safety, and significantly enhanced accessibility for particular groups of people that find using traditional modes difficult (mobility impaired people, those in highly remote, rural locations for example), there is as yet very little articulation of what the benefits of the smart transition are at the level of society as a whole.

Perhaps the time has come for a twenty-first century Jane Jacobs figure to pose the simple yet exceedingly difficult to answer question of 'why are we doing this?' to the smart era's equivalents of Robert E Moses. It took decades from Jacobs' pivotal intervention for her challenge to enter the mainstream policy debate about how we manage mobility in our cities. The dash to implement smart mobility that we see now is highly reminiscent of the clamour to give over first space and then whole places to the automobile. Although it is undeniable that the car changed many things for the better, many places now rue the decision to give it free rein for so long. We are at a similar juncture now with smart technology: if we do not actively shape it, it will shape our future lives for us. The critical imperative is, therefore, to use whatever governance capacity we have in the first instance to debate and decide what kind of mobility future we want to try and facilitate while we are able to do so.

## REFERENCES

Bilger, G. (2013). Auto correct. *The New Yorker*, November 25. Retrieved from https://www.newyorker.com/magazine/2013/11/25/auto-correct

Capoccia, G., & Kelemen, D. (2007). The study of critical junctures: Theory, narrative and counterfactuals in historical institutionalism. *World Politics*, *59*(3), 341–369.

Docherty, I., & Shaw, J. (2011). The transformation of transport policy in Great Britain? 'New Realism' and New Labour's decade of displacement activity. *Environment and Planning A*, *43*(1), 224–251.

Dowling, G. (2014). The curious case of corporate tax avoidance: Is it socially irresponsible? *Journal of Business Ethics*, *124*, 173–184.

Fagnant, D., & Kockelman, K. (2015). Preparing a nation for autonomous vehicles: Opportunities, barriers and policy recommendations. *Transportation Research Part A: Policy and Practice*, *77*, 167–181.

Giddens, A. (2008). *The politics of climate change*. London: Policy Network.

Gray, D., Docherty, I., & Laing, R. (2017). Delivering lower carbon urban transport choices: European ambition meets the reality of institutional (mis)alignment. *Environment and Planning A*, *49*(1), 226–242.

International Transport Forum Corporate Partnership Board. (2017). Transition to shared mobility: How large cities can deliver inclusive transport services. *OECD-ITF Corporate Partnership Board Policy Insights*, June.

Lindberg, G., & Fridstrøm, L. (2015). *Policy strategies for vehicle electrification*. ITF/OECD Discussion Paper 2015:16. Paris: ITF/OECD.

Marsden, G., & McDonald, N. (2017). Institutional issues in planning for more uncertain futures. *Transportation*, online advance publication.

Marsden, G., & Stead, D. (2011). Policy transfer and learning in the field of transport: A review of concepts and evidence. *Transport Policy*, *18*(3), 492–500.

Meadowcroft, J. (2009). What about the politics? Sustainable development, transition management, and long term energy transitions. *Policy Sciences*, *42*, 323–340.

Ministry of Transport. (1963). *Traffic in towns*. London: Ministry of Transport.

Paget-Seekins, L., & Tironi, M. (2016). The publicness of public transport: The changing nature of public transport in Latin American cities. *Transport Policy*, *49*, 176–183.

Sadik-Khan, J., & Solomonov, S. (2016). *Street fight*. New York, NY: Viking.

Schöller-Schwedes, O. (2010). The failure of integrated transport policy in Germany: A historical perspective. *Journal of Transport Geography*, *18*(1), 85–96.

Wockatz, P., & Schartau, P. (2015). *IM traveller needs and UK capability study: Supporting the realisation of intelligent mobility in the UK*. Milton Keynes: Transport Systems Catapult.

# CHAPTER 3

# THE CASE OF MOBILITY AS A SERVICE: A CRITICAL REFLECTION ON CHALLENGES FOR URBAN TRANSPORT AND MOBILITY GOVERNANCE

Kate Pangbourne, Dominic Stead, Miloš Mladenović and Dimitris Milakis

ABSTRACT

*This chapter provides a reflective critique of Mobility as a Service (MaaS), an emerging development seeking a role within the Smart Mobility paradigm. We assess a range of its future implications for urban policymakers in terms of governance and sustainability (i.e., social and environmental impacts). We begin by describing the origins of the MaaS concept, along with the features of precursor technologies and current early examples. We then reflect on the marketing of MaaS and use it to consider how we might anticipate some potentially less desirable aspects of the promoted business models. Finally, we discuss the implications for governance.*

**Keywords:** Smart mobility; governance; sustainability; Mobility as a Service

## INTRODUCTION

Urban governance is experiencing significant challenges, not least an accelerating shift from public to private provision, sometimes associated with austerity policies. At the same time, a scalar shift for transport governance is seen in the trend for devolution of responsibility to institutions at local and regional levels. We are also witnessing significant shifts in social practices and expectations, facilitated by advanced mobile information and communications technologies (ICTs), an increased dependence on online service provision and a rise in demand for and supply of 'flexibility' in the provision of various types of service. Simultaneously, there is an ongoing and urgent need for the transport sector to address local and global problems that it plays a significant role in creating: urban congestion, noise, air pollution, public health, transport safety, unequal access to services and climate change emissions (Stead, 2016).

In this chapter we illustrate our account of this urban and mobility governance challenge by analysing one 'spearhead effort' that is generally referred to as Mobility as a Service (MaaS), a recent concept in the 'Smart Mobility' arena. MaaS represents a hybrid innovation, as a platform technology combined with a business model for delivering integrated access to transport services. This is sometimes termed a 'multi-sided platform'. As a tool for integration, it can, in principle, incorporate all currently available transport modes as well as emerging technologies such as self-driving vehicles, and as such can make it a tempting development for public authorities. Moreover, the MaaS concept could expand to include urban logistics and other services (e.g., gym, cinema or restaurant bookings), integrating these with the transport service needed to access them. However, underpinning the apps and the packages offered to users, there are business models. The choice of business model and the detailed design of the value offer is not trivial, raising important questions about inclusiveness and sustainability, potentially threatening the common good. Given these questions, we highlight the risks to achieving a more sustainable transport system through the commodification of access to mobility by commercial intermediaries who provide ICT-based aggregation services to both end users and transport service providers. Furthermore, we unpack these issues to address the question of what MaaS might mean for the governance of mobility and urban development. Due to limited space we are unable to broaden the analysis to other critical issues such as the risk of mobility enclosure and its impact on the human right for freedom of movement, or the details of market regulation (for example in relation to acting as a 'reseller' for transport tickets) and consumer protection (in relation to both data and transport service levels), but we acknowledge that these issues are also of significant interest.

## CONSTRUCTION OF MOBILITY AS A SERVICE

As a recent mobility concept, the definition of what is, or is not, MaaS is not fully solidified (see Chapter 4 for a review of why definitions matter). Jittrapirom et al. (2017) review 12 conceptualizations, classifying a set of core MaaS characteristics. These characteristics include the integration of transport modes, tariff options, a single platform, multiple actors, use of technologies, registration requirement and a user-centred orientation with personalization and customization. Overall, MaaS tends to consist of a platform that integrates access to information about and payment for multiple combinations of transport services.

With these characteristics in mind, we briefly describe the emergence and early development of MaaS concepts drawing from early experiences in Finland, the Netherlands and the United Kingdom, three countries where a more fully realized version of the concept is seeking to challenge the current landscape of urban transport provision. The MaaS discourse is driven by business and technology priorities. Hoadley (2017) describes how lobbying from the digital and intelligent transport systems industries, supported by innovators in the personal transport sector (e.g., car sharing), is influencing policy thinking at higher scales. However, as the involvement of the public transport sector in MaaS has been limited, wider evidence at city/regional levels is missing.

The central assumption of MaaS as promoted by MaaS Global (often credited with inventing the concept) is that transport services can be converted into service packages, as with the telecommunications sector. Registered users can select a package that bundles access to several transport modes, ranging from conventional public transport to taxis (shared or solo) and vehicle sharing such as cars, scooters and bicycles (electric or otherwise). The monthly number of trips by each mode is determined by packages purchased on the basis of the user's expectation regarding the number of trips s/he needs. The expected result is provision of door-to-door mobility services, with the promise of greater 'efficiency' and the opportunity to break car-dependency. The benefits of bundling various transport services together through one digital interface are presented as good for both the customer and the operator. Wherever elements of MaaS are being rolled out under the Smart Mobility banner, we are offered variants of the same dream: 'seamless and effortless' (MaasAllianceEU), 'Smarter, Faster, Greener', 'on-demand tailor-made transport' and 'instant Access' (MaaS Scotland).

The vision, as is overwhelmingly clear in the promotional rhetoric, dominantly focuses on envisioning 'positive' effects. From the perspective of a multi-level model of innovation, this rhetoric is an inevitable process of niche actors challenging the incumbent actors of the regime they are targeting (Shove & Walker, 2007). This is resulting in promotional alliances, such as the European MaaS Alliance (http://maas-alliance.eu) and Scotland's MaaS Scotland (https://maas-scotland.com/), bringing together public and private actors who have been engaged in concept formation with a number of prototypes, building and converging towards the first emergence of the term itself in Finland (Heikkilä, 2014). The prototypes exist on a continuum starting from what might be termed 'pre' or 'partial MaaS' such as Smartcard-based integrated ticketing systems to full-integrated apps. In addition, the European Union has funded several pilot projects that focus on different services or technologies that collectively have informed the development of the higher level of integration conceptualized for MaaS. For example, as part of Superhub, a MaaS-type approach was used to try and incentivize people to use sustainable transport.[1]

Developments from the sharing economy have informed the development of MaaS. Many peer-to-peer products to share vehicles, traffic information or offer rides have been developed, such as Flinc, Waze, Car2Go,

BlaBlaCar, ReachNow, Zipcar, CoWheels and Faxi, many of which can in principle be accessed through a MaaS platform (Flinc was part of the Berlin field test of MyWay). However, MaaS is not a necessary development for most of the new transport technologies, such as electric or self-driving vehicles. Ultimately, in order to appeal to users *and* to deliver on the objectives of local authorities, the full MaaS vision needs a unique selling point. This is the implicit promise of making more efficient use of diverse transport services by simplifying access to them in more flexible combinations, while emphasizing the sharing of mobility resources, thus speaking both to the policy objectives of mobility governance and of a better and cheaper mobility offer for users. This is a key claim of the field leader, MaaS Global, for its product, Whim.

The imminent introduction of Whim in different national contexts (Finland, the United Kingdom, the Netherlands and Singapore) suggests that concept transfer of 'full MaaS' is already occurring through concerted effort on the part of MaaS Global. These efforts can engage with actors at city, regional, national and supra-national scales, to offer more 'efficient' door-to-door mobility and promote 'sustainable urban transport'. However, there are various unresolved societal issues in deploying commercial MaaS in real life, some of which are discussed below.

## Complexities and Contradictions in MaaS for Real

Transport professionals are well aware of the history of unanticipated consequences around many twentieth-century mobility innovations such as biofuels (Morton et al., 2017). Therefore, in this section we now consider two interdependent issues with MaaS that are under-examined. Based on existing deployments of MaaS, we consider that these could be predicted as having undesirable consequences. First is the choice of the business model, its formulation and possible impact on aspects of mobility resilience. Second is the promise of freedom, a central component of the MaaS rhetoric.

### *Business Models and Resilience*

The Multi-Sided Platform nature of MaaS is critical to its value proposition: the benefits to users on different sides accrue as users *on every side* increase in number, increasing the opportunities for interactions and subsequent revenue generation (Jittrapirom et al., 2017). Gaining a critical mass of MaaS users to both demand and supply services is crucial to success, as highlighted by Finger, Bert and Kupfer (2015) and Sochor, Strömberg and Karlsson (2015). The business objectives are also an essential element in whether a MaaS has potential to achieve social or environmental benefits (e.g., through stimulating beneficial behaviour change). Some of the business models of early Smart Mobility entrants are aggressive (e.g., Uber), and many disrupt existing provision (e.g., Uber disrupts the traditional taxi and private hire markets, and Obike (and other) dockless bikes have challenged both city authorities' control of their jurisdiction and existing dock-based bike-share systems). It is not clear that all the providers are looking to be part of an aggregation model, in much the same way that, where competition is allowed for bus routes, the outcome is not integration.

Operators of shared services are increasingly being relied upon as a substitute for public transport in some jurisdictions (notably the United States). For example, in Florida, some administrations subsidize residents' Uber trips instead of extending bus routes. In Altamonte Springs public transport has been drastically reduced, and all Uber trips are subsidized by at least 20%. In turn, this could affect mobility resilience, given that Uber has released audited accounts that show that it is making large losses despite high turnover (*Financial Times*, 2017). Given the use of aggressive customer subsidies to build the business, there are two important concerns. One must be the risk of business failure which would leave car-less residents without mobility due to the reduction in support for socially necessary services. This is a risk even if a ride-hailing service is integrated into a MaaS. If the aggregator has no alternative provider for the journeys that the ride-hailing service supplied, then those customers who were reliant on it have no mobility. Whether this is the case may depend on whether or not certain operators demand (or are offered) exclusivity within the MaaS product. The second concern is what happens when most Uber users must pay the full, rather than subsidized, cost of their journeys and there is no longer a public transport alternative (Lee, 2016) as ride-sharing has been shown to reduce use of public transport (Clewlow & Mishra, 2017).

Another question is MaaS's reliance on registration and digitalization, which create additional barriers for those who are already experiencing exclusion, adding a loss of mobility to problems caused by, for example, the digital gap or through lack of access to banking. This is an important question, since transport operators increasingly offer 'discounts' to smart-card and app users, with those using traditional payment methods, such as cash at the point of use, paying more for the service.

Finally, there has been little discussion on the vulnerability in relation to MaaS's dependence on ICT. There is the potential for an entire city to come to a standstill, should the MaaS system be compromised, for example through power failure, ICT failure or a Deliberate Denial of Service cyberattack. The transport sector is a critical infrastructure, having been the focus (or means) of criminal and terrorist attacks (Theoharidou, Kandias & Gritzalis, 2012). While the dangers of a compromised MaaS system may not be as serious as say a cyberattack on a fleet of self-driving vehicles, its disruption potential on urban mobility is still substantial, suggesting that if access to transport is mediated via MaaS platforms, these clearly need to be included in Critical Infrastructure Protection strategies.

*The False Promise of Freedom*

Selling of the notion of 'freedom' in the context of a finite transport network and environmental limits raises the need to have a debate about individual and collective rights/responsibilities. MaaS Global advertises itself as 'mobility on a whim', promoting an ideal of individual unfettered freedom. This promise is at odds with the challenge of satisfying simultaneous demand in a finite transport network. By drawing parallels for MaaS packages with those used in telecoms or media streaming, the impression is given that any desired trip can be made at any time (any origin to any destination). However, telecommunications and transport networks have different network capacity properties. ICT network capacity is more easily scaled-up as network demand is managed through data package routing protocols, as data ascribes no emotional or economic value to its path from origin to destination. Telecommunications network congestion can be managed in ways that are impossible in a transport network. Data can be prioritized, held in a buffer, or rerouted through different nodes, not necessarily the shortest path. This is not the case for humans moving through urban transport systems, who will know if they are deliberatively delayed or diverted, and will complain, or even rebel. Thus, it is hard to see how MaaS can deliver its promise of freedom through its packages of different levels of pre-purchased or 'Pay-As-You-Go', if the network is at capacity at the point at which a customer requests service.

The promise of freedom also fails to acknowledge that current problems of traffic congestion, urban air pollution and greenhouse gas emissions arise from the aggregate impact of our individual activities. In the drive to develop a customer base, MaaS could feed unsustainable individual practices rather than restraining and redirecting people to more sustainable transport modes. Hoadley (2017) highlights this risk, citing the 'poor visibility given to public transport in current MaaS discussions and developments' (p. 7). The same is potentially true for non-motorized transport modes (walking and cycling) which may be sidelined as mobility options because they do not generate substantial income for MaaS (see also Chapter 2 for further discussion on unaligned commercial behaviours).

Furthermore, the potential for a rebound effect, where energy (time or travel) savings in one area are 'cashed-in' by increasing use in another area, is largely ignored (see also Herring & Sorrell, 2008). This is glossed over in promotional scenarios:

> *After a month of using MaaS, Melinda's family life has completely changed. They have sold Melinda's car and offer the other car for short term rental using the MaaS operator's website (community car club). In exchange Melinda's family gets credit in their MaaS account, which they use to buy mobility services. (TSC, 2016)*

In this example, Melinda's family have become totally dependent on the MaaS service for all their mobility, and have been able to make time and cost 'savings'. However, the money they accrue by renting out their remaining vehicle is limited to use for other mobility services. Thus, MaaS is able to frame their mobility practices, by making the offer of credit that can only be spent within the MaaS system. Should a package allow six taxi trips per month, for example, the theory of loss aversion (human cognition is more attuned to avoiding a loss than achieving a gain) (Tversky & Kahneman, 1991) would suggest that users will experience regret if they do not 'use up' their trip allowance, potentially resulting in induced trips adding further pressure to the system. This means that the design of packages and pricing is crucial – users should be allowed to 'roll-over' unused credits to prevent the risk of induced travel. There is some parallel to the mobile telecommunications market, where Gerpott and Thomas (2014) have shown tariff-type impacts on consumers' data usage intensity. Industry research also suggests that mobile customers buy more expensive packages than they need to avoid the risk of paying high charges for going over their data allowances. Since early 2016 in the UK 'Data Rollover' packages have started to be introduced as a result (uSwitch, 2017).

## IMPLICATIONS FOR GOVERNANCE

In this section, we consider the governance implications of the issues highlighted in the preceding sections and raise some key questions, in order to highlight where stakeholders might be advised to take particular care before making MaaS a central pillar around which urban transport is organized. In several respects, MaaS repackages existing Intelligent Transport System ideas of integration, and sounds intuitive, understandable and attractive, in part because achieving 'seamlessness' between modes has been a goal of transport authorities for many years. As MaaS is promoted in this way, it is hard for stakeholders not to embrace it. This positive framing is a clear attempt to win a significant place in the market by MaaS providers, who are primarily private companies. However, the enthusiasm with which organizations are embracing the concept masks some significant uncertainties around governance in relation to control and setting strategic goals. While MaaS has developed with little direct public steering (MaaS International, 2017), in Finland, the Ministry of Transport has been quite engaged in providing support to help MaaS into the marketplace (Finnish Transport Agency, 2015), with various incentives and a hope for 'another Nokia' (i.e., supporting innovation with money but little regulation, in keeping with a free-market ideology). However, it is starting to be recognized that this will need to change, as the steering efforts so far have not taken into account the full extent of complexity and unanticipated consequences from MaaS. Clearly, there are different roles that public authorities can take (e.g., enablement, leadership, laissez-faire). Four key choices are set out below.

First, decision-makers need to be able to assess and compare transport systems/infrastructure investments/policies, but this poses a number of challenges. The greatest issue is the inevitable uncertainty about the direction of technological development and its impacts. One way of addressing this uncertainty is to create a controlled but open structure for research and testing.

Second, there needs to be a process for negotiating and ascribing liabilities across a complex web of stakeholders, addressing consumer protection, developing market rules and defining the role of the public sector. For example, there is a need to set minimum service-standards to protect socially necessary services, or ensuring that the cheapest public transport fares are available to MaaS users, however small their service use may be.

Third, it is important to highlight that the MaaS concept includes a need for a set of organizations, legislation and other aspects that collectively serve to lock a technology into society. MaaS is a technological assemblage and not solely an App, the value concept of service packages or the revenue streams that define the business model. MaaS's uniqueness is the potential to involve so many different individual technologies, both ICT and transport, and the ability to position the concept as an optimizer. For it to work, it requires the MaaS operators to occupy a very powerful place in the network both in a co-ordinating space and a price-setting space. This is something that has proved almost impossible for city-led transport systems in the past. It is difficult to imagine such a powerful position in the governance network being easily obtained, but if it is, then it would certainly need regulating.

Finally, there are risks of inaction by the state because doing nothing is not the same as no change, as this is already occurring. As models such as Uber or Lyft have made taking a taxi exceptionally convenient, there is evidence that this reduces public transport use: a taxi can provide door-to-door service, and where more than one ride-hailing service is present, price competition results in pulling custom away from transit services (Clewlow & Mishra, 2017; Sadowsky & Nelson, 2017). MaaS also introduces a commercial intermediary between citizens and public transport providers, diluting brand image (Hoadley, 2017), suggesting a reason why it has taken longer for MaaS platforms to engage with long-established public transport operators. For example, in the West Midlands, Whim has successfully negotiated Gett taxis, National Express buses and Midland Metro trams, the regional city bike and rental car providers, but has not yet persuaded other large public transport providers such as FirstBus or Arriva. Thus, relying on MaaS to relieve cities from car dependency and related congestion is an outcome that could be further undermined if the door-to-door convenience of ride-hailing (and eventually self-driving vehicles) becomes a reality without strong steering by the state and forward-thinking strategies in place to address these conflicting forces.

There is a need for strategic thinking about urban technology, as the integration of the built environment, hard infrastructures and digital services. However, no cities have yet incorporated MaaS into transport, environment and energy policies (Li, 2017), even though quite large elements of MaaS are operating in several places, as described above. This is a strategic omission that reduces the opportunity for MaaS to be designed to contribute to sustainable urban mobility. According to Li (2017), this is because everyone believes in the idea that MaaS will *automatically* contribute to sustainable urban mobility through 'efficiency'. However, citizens and governance actors need to be able to decide which modes should be prioritized according to the social and environmental needs of their jurisdiction, and specify MaaS packages accordingly. However, it is striking that both Li (2017) and Hoadley (2017) note the lack of engagement from city and regional authorities at this stage. For example, the UK Transport Systems Catapult did not identify a role for transport authorities in the MaaS ecosystem beyond being a 'customer' for data (TSC, 2016). This dominance of producer-led visions is also a feature of autonomous vehicle innovations as discussed in Chapter 5.

The current focus on outsourcing innovation to the private sector combined with the competitive national rhetoric predicated on economic growth through mobility innovations suggests that government may be tempted to cede control of outcomes to market forces. This path carries profound implications for decision-making in transport and urban governance, as there is a critical governance gap in relation to managing the Smart Mobility transition if regulation is removed in a bid to placate private-sector demands. MaaS innovators are primarily private-sector firms who are attempting to steer the development of the mobility system in ways that serve their vested interests (Vergregt & Brown, 2007), and regulatory capture through manipulating transport governance mechanisms does have a precedent (Morton et al., 2017). The further commodification of urban mobility, while offering opportunities to some consumers, is not synonymous with being able to steer mobility systems to more desirable outcomes. However, there are models where the public sector remains at the heart of the system if not the technologies.

In summary, some governance levers could be lost through ideological pressure to create revenue streams out of previously public goods, endangering the achievement of social and environmental goals that are intertwined with mobility provision. While recognizing the positive potential of MaaS, it should not be presumed to deliver a uniquely positive set of outcomes for all. Strategic management is needed to set objectives, monitor mode share changes and to understand social, distributional and environmental impacts, as well as to provide an environment where innovation (by both the public and the private sectors) can flourish. Risks also need to be addressed, in order to understand whether the transport efficiency gains that might be realized through the wholesale adoption of MaaS are jeopardized by a resilience gap.

## CONCLUSIONS

In this chapter, we have highlighted that MaaS represents a conceptual approach to delivering service to users that is not a fixed product. Conceptual elements exist through individualized services, but MaaS is increasingly promoted as an integrated product capable of shaping how transport is organized and managed in cities. We have illustrated this point through a short description of the construction of the MaaS concept and given an account of its emergence in early sites of innovation for integrated MaaS platforms.

We have highlighted the risks posed by the business models to meeting key policy aims such as congestion reduction and climate change mitigation, as well as touching on the social inclusion aspects. We have also highlighted the potential threat to transport and social resilience through over reliance on single operators of innovative services, and the potential effects of innovative services on existing services. The result could be a deepening of exclusion by over digitalizing and enclosing access to transport services and through cyberattack vulnerability.

The dominant rhetoric surrounding MaaS is technologically deterministic and highly optimistic. However, we contend that advertising MaaS as 'mobility on a whim' promotes a false promise of individual unfettered freedom that fails to acknowledge that current problems of traffic congestion, air and noise pollution, and greenhouse gas emissions are large-scale problems arising from the aggregate impact of our individual activities, with a wide distribution of mobility habits.

While MaaS could be designed to influence behaviours to be more sustainable, the commodification of mobility through the service package approach requires customers who buy services. Profitability of private businesses inevitably requires the use of these services. Thus, MaaS has a strong potential to result in *increased* mobility amongst those who can pay for it (and have paid in advance). Steering MaaS developments towards more desirable and inclusive societal outcomes requires engagement by the state in the design goals, pricing structures (and subsidies), coverage and consumer protection. In so doing, it may offer opportunities to overcome longstanding challenges to truly integrated transport services. The alternative path where MaaS is seen as the solution through outsourcing the challenge of mobility co-ordination and where the state shrinks in its capacity to co-ordinate and steer seems fraught with risks that would be difficult to reverse. However, the situation is not one of a dichotomy between the opposing paths of laissez-faire and state-led regulation, though there is a need to avoid possibly damaging technology lock-ins. While the technology is in its foundational development stage, there is an opportunity to address the consumer issues in a proactive or even participatory way by stimulating debate about the proper role of the state in addressing citizens' fundamental mobility needs.

## NOTE

1. Many such initiatives are documented by Jittrapirom et al. (2017) and Kamargianni, Li, Matyas, House, and Count (2016), including UbiGo, Smile, Tuup, Moovel and Whim.

## REFERENCES

Clewlow, R. R., & Mishra, G. S. (2017). *Disruptive transportation: The adoption, utilization and impacts of ride-hailing in the United States*. Research Report UCD-ITS-RR-17-07. Institute of Transportation Studies, University of California, Davis.

*Financial Times*. (2017). Uber recorded a $2.8bn loss in 2016 in the middle of an aggressive global expansion, cementing its place as the most heavily-lossmaking private company in the history of Silicon Valley. Retrieved from https://www.ft.com/content/52b54056-214d-11e7-b7d3-163f5a7f229c. Accessed on 16 October 2017.

Finger, M., Bert, N., & Kupfer, D. (2015). *3rd European intermodal transport regulation summary 'Mobility-as-a-service: From the Helsinki experiment to a European model?'* Technical Report, European Transport Regulation Observer No 2015/01.

Finnish Transport Agency. (2015). MaaS services and business opportunities. Retrieved from http://www2.liikennevirasto.fi/julkaisut/pdf8/lts_2015-56_maas_services_web.pdf

Gerpott, T. J., & Thomas, S. (2014). Empirical research on mobile Internet usage: A meta-analysis of the literature. *Telecommunications Policy*, *38*, 291–310.

Heikkilä, S. (2014). Mobility as a service: A proposal for action for the public administration, Case Helsinki. Retrieved from https://aaltodoc.aalto.fi/bitstream/handle/123456789/13133/master_Heikkil%C3%A4_Sonja_2014.pdf?sequence=1

Herring, H., & Sorrell, S. (Eds.). (2008). *Energy efficiency and sustainable consumption: The rebound effect*. Basingstoke: Palgrave Macmillan.

Hoadley, S. (Ed.). (2017). *Mobility as a service: Implications for urban and regional transport*. Discussion paper. POLIS, Brussels, Belgium.

Jittrapirom, P., Caiati, V., Feneri, A.-M., Ebrahimigharehbaghi, S., Alonso-González, M. J., & Narayan, J. (2017). Mobility as a service: A critical review of definitions, assessments of schemes, and key challenges. *Urban Planning*, 2, 13–25.

Kamargianni, M., Li, W., Matyas, M., House, C., & Count, W. (2016). A comprehensive review of 'mobility as a service' systems. In *95th Annual Meeting of the Transportation Research Board*. Washington DC.

Lee, D. (2016). Is Uber getting too vital to fail? Retrieved from http://www.bbc.com/news/technology-38252405. Accessed on 28 August 2017.

Li, Y. (date 2017). *Future roles of public authorities in mobility as a service (MaaS)*. Workshop Report. Smart Procurement for Better Transport H2020 Project.

MaaS International. (2017). Finnish company MaaS Global completes funding round, raising €14.2 million. Press Release 02.08.2017. Retrieved from http://maas.global/press/. Accessed on 30 August 2017.

Morton, C., Budd, T. M., Harrison, G., & Mattioli, G. (2017). Exploring the expectations of transport professionals concerning the future automobility system: Visions, challenges, and transitions. *International Journal of Sustainable Transportation*, *11*, 493–506.

Sadowsky, N., & Nelson, E. (2017). *The impact of ride-hailing services on public transportation use: A discontinuity regression analysis*. Economics Department Working Paper Series 13. Retrieved from http://digitalcommons.bowdoin.edu/econpapers/13

Shove, E., & Walker, G. (2007). CAUTION! Transitions ahead: Politics, practice, and sustainable transition management. *Environment and Planning A*, *39*, 763–770.

Sochor, J., Strömberg, H., & Karlsson, I. C. M. (2015). Implementing mobility as a service challenges in integrating user, commercial, and societal perspectives. *Transportation Research Record*, *4*, 1–9.

Stead, D. (2016). Key research themes on governance and sustainable urban mobility. *International Journal of Sustainable Transportation*, *10*(1), 40–48.

Theoharidou, M., Kandias, M., & Gritzalis, D. (2012). Securing transportation–Critical infrastructures: Trends and perspectives. In C. K. Georgiadis, H. Jahankhani, E. Pimenidis, R. Bashroush, & A. Al-Nemrat (Eds.), *Global security, safety and sustainability & e-democracy: Lecture notes of the Institute for Computer Sciences, Social Informatics and Telecommunications Engineering* (Vol. 99). Berlin, Heidelberg: Springer.

Transport Systems Catapult. (2016). Mobility as a service: Exploring the opportunity for mobility as a service in the UK.

Tversky, A., & Kahneman, D. (1991). Loss aversion in riskless choice: A reference-dependent model. *The quarterly journal of economics*, *106*(4), 1039–1061.

uSwitch. (2017). What is data rollover and which networks offer it? Retrieved from https://www.uswitch.com/mobiles/guides/what-is-data-roll-over-and-which-networks-offer-it/. Accessed on 25 October 2017.

Vergregt, P. J., & Brown, H. S. (2007). Sustainable mobility: From technological innovation to societal learning. *Journal of Cleaner Production*, *15*, 1104–1115.

# SECTION TWO

# WHOSE VOICES ARE IN THE SMART MOBILITY DEBATE?

# CHAPTER 4

# SMART MOBILITY: DISRUPTING TRANSPORT GOVERNANCE?

Robyn Dowling

ABSTRACT

*Transport governance is dependent on, and works through, legally defined or socially accepted categories that are challenged by the technological and business disruptions that characterize smart mobility. This chapter explores the dynamic interactions between the categories of transport governance and new (disruptive) forms of mobility, using a framework that highlights the ways governance solutions are influenced by how problems are framed. The argument is made through two empirical cases – of car sharing and motorized (electric) personal mobility devices. Car sharing emerged quietly onto the transport landscape in Australia, and has been accommodated and facilitated by local government parking policies. In this case, the tools for governing smart mobility already existed, and became adaptable to new imperatives. Personal mobility devices (PMDs), battery-powered motorized devices designed to be used by an individual on footpaths or shared user paths, are neither common nor legal on many roads or footpaths. The process through which PMDs became regulated in the Australian state of Queensland is used to illustrate the ways in which 'epistemic experimentation' can challenge regulatory framings and foster the introduction of new forms of smart mobility. The chapter concludes that smart mobility both disrupts and confirms dominant framings of transport governance, especially in relation to automobility, and that further challenges will need to be met as autonomous vehicles become more widespread across transport infrastructure.*

**Keywords:** First-and-last mile; electric mobility; car-sharing; technological disruption; framing

## INTRODUCTION

Transport governance is dependent on, and works through, legally-defined or socially accepted categories. In this chapter I am concerned with the challenges smart mobility poses to these categories and framings. For example, there is necessity to rewrite road and driver licensing rules in order for autonomous vehicles to legally operate on roads (Tranter, 2017), while new forms of transport such as that offered by Uber have resisted regulation as a form of transport (Dudley, Banister, & Schwanen, 2017). Mobility-on-demand or Mobility as a Service (MaaS) models are neither mass transit nor individual transit, but a blurring of the two, and as such present the general challenges of governing the sharing economy (see Gurran & Phibbs, 2017 and extended discussion in Chapter 3). In this chapter I explore the dynamic interactions between the categories of transport governance and the new (disruptive) forms of mobility they seek to govern. I do so through two empirical cases – of car sharing and motorized (electric) personal mobility devices. The chapter begins with an elaboration of the conceptual framework, and in particular the ways hegemonic automobility is shaping options for governing smart mobility. I then turn to the two cases, illustrating the differing ways in which existing and novel categorizations have shaped the introduction and growth of these new forms of mobility into existing transport conclusions. In the conclusion I reflect on the implications of the argument for more radical forms of smart mobility such as autonomous vehicles.

## AUTOMOBILITY, CATEGORIES AND FRAMING IN TRANSPORT GOVERNANCE

My take on governing smart mobility emerges out of a broad definition of governance as not just policy, and not just actions of the state (Dowling & Kent, 2015). Rather, governing occurs through a process in which problems are identified, and interventions formulated and implemented that seek to produce desired outcomes and avert undesired ones (Bulkeley, Powells & Bell, 2016). Categories and framing, by which I mean the links between the ways a problem is defined and the solutions proposed, are critical to these governance processes (Lawhon & Murphy, 2011). For example, transport policies and regulatory frameworks are discursively framed: constructed and articulated through norms, values and beliefs (Aldred & Tepe, 2011; Curtis & Low, 2012). In turn, these frames inform the possibilities of governance: what is governed and how it is governed.

Important in Australian cities (and many others worldwide) is the frame of 'automobility', conceived as a configuration 'of political interests, economic determinants, cultural values, physical infrastructures and technological capacities' (Cohen, 2012, p. 379) in which car-based movements are hegemonic (Henderson, 2009). In broad terms a wide-range of transport governance mechanisms privilege car-based urban infrastructures and hence further sustain automobility (Brown, Morris & Taylor, 2009). These impact the ways in which technological disruptions enter the transport system. An analysis of planning for electric mobility in Iceland, for example, identifies a continuation of transport engineering approaches to the city that may bolster rather than dismantle hegemonic automobility (Driscoll et al., 2012) while Chapter 5 discusses this in the context of autonomous vehicles. Automobility is also embedded in transport infrastructures. Roads, bikeways, footpaths, shipping channels and the like shape urban mobility by channelling the flows of people and transport objects across cities and in facilitating car dependency (Shove et al., 2015). Infrastructures are both regulated and political: there is considerable debate on what types of mobility are appropriate for what types of transport spaces. For example, Bonham and Cox's Foucauldian analysis of cycling policy suggests that 'the transport rationalization of movement prioritizes and allocates space according to speed and order so that practices of walking and cycling become difficult to place on the modern street with cyclists' use of road space contested and marginalized' (Bonham &

Cox, 2010, p. 50). Likewise, the allocation of road space to parking for commercially operated car share continues to be hotly contested (Dowling & Kent, 2015). In sum, transport infrastructure in Australian and many other cities is characterized by the privileging of automobile-based flows and the accompanying frictions for other slower and smaller means of mobility.

Just as infrastructures are obdurate, there is a path dependency to transport governance. In other words, the categories through which governance is thought and enacted (Blomley, 2014) shape and limit futures. Potter and Enoch's description of paratransit is most apt here:

> *The problem is that, almost by definition, paratransit alternatives do not comfortably fit under any of these categorisations with the result that they have no institutional home and either upset the status quo (as with Uber currently) or else are stillborn. (Potter & Enoch, 2016, p. 28)*

New and disruptive transport businesses and modes that characterize smart mobility, such as paratransit, are out of alignment with the predominant categories of transport governance.

Yet, as Blomley (2014) also reminds us, such categories are always fissured and open to contestation: there is potential for smart mobility governance to move beyond the frames of the present. This is already in evidence with the case of Uber which refuses to be defined as a taxi service. Conceptually, it is also possible that new alignments of thought, problem definition and techniques of governance may enable the governance of smart mobility beyond the frame of the present. Just as automobility is fissured (Dowling & Simpson, 2013) so too are categories and regulation; categories, law and regulations are adaptable and constantly in process. There is a fuzziness in the implementation and interpretation of legal categorizations in practice (Blomley, 2014). Moreover, governance is constituted by myriad small and mundane decisions which are dynamic rather than static, and as such can be important bedrocks of change, even when they are not characterized as disruptive (Haarstad, 2016). Finally, inspiring new framings (imaginaries) of what is possible and the thinking of new categories can enliven mobility governance. 'Epistemic experimentation' – testing new ways of knowing a phenomenon – can foster new ways of governing it (Jensen, Cashmore & Elle, 2017). For example, the shift of cycling from a road safety problem to an experience and as a mode to be accounted for like other forms of traffic underpins new forms of cycling governance and practice (Jensen et al., 2017). I consider both of these ways in which smart mobility is reframing transport and its governance categories, using car-sharing as an example of a small-scale shifting, and personal mobility devices as an instance of epistemic experimentation.

## CAR SHARING

My first example is car sharing (known as car clubs in the United Kingdom), one of the earliest, and least confronting elements of smart mobility. Car sharing has been phenomenally successful in automobile-dominated Sydney, in 15 years growing from two cars and 10 members to over 2,000 cars and 10,000 members (Kent, Dowling & Maalsen, 2017). Car sharing is a complex ensemble of materials and infrastructures, both smart and otherwise. It is dependent on a vehicle that is accessed for use rather than owned. The car is also both part of a field of, and dependent on, objects such as roads, gasoline, traffic signals, car parking spaces and pedestrians. Like other sharing economies, car sharing is digital, facilitated by and positioned in computing, the Internet and web 2.0 (Belk, 2014). Car sharing encompasses diverse digital systems – accessed by a smart card, booked and paid for electronically, user tracked and calculated via GPS. Car sharing emerged quietly on to the transport landscape, largely under the transport policy radar (Kent & Dowling, 2016). But car sharing's growth was built on policy foundations, just not conventional transport pillars. I explore these foundations in this section as a means of illustrating the ways in which smart mobility can be enabled by existing mechanisms of transport governance.

In Sydney, car sharing has been supported through subtle modifications to existing land use and other planning mechanisms, all focused on parking. This is for two reasons. First, car parking, and the allocation of car parking spaces, is most often the purview of local land use planners (Kent & Dowling, 2016; Shoup, 2014), and it is at the local level that most innovation and attention to sustainable transport has occurred (McGuirk et al., 2015). Second, car sharing relies on the guarantee of a car parking space. Allocated parking space works to provide consistency and reliability in the location of the act of transfer where the car is conferred from one user to the next. Fixed and reserved parking allocations relieve car-sharing organizations of the need to invest in parking, and contribute to the financial viability of the business, particularly in the start-up phase.

All local government areas in Sydney which have car sharing also have car-sharing policies. These policies do not contain incentives for residents to take up car sharing, or advertising campaigns to raise awareness. Nor do they include the use of car sharing for local authority vehicle fleets (as happens elsewhere such as the United Kingdom), though this might be the most obvious element of such a policy. Instead, these car-sharing policies are principally about the allocation of parking. In general, the car-share policy is used to articulate a local authority's 'bequest' of parking for car share cars, and the conditions of the bequest. These conditions include payment of a fee, responsibilities of the car-share operator to provide data and annual reporting requirements, and quite complicated rules about the location of a car-share parking space, such as preferably not in front of a private residence but near public spaces such as parks or council offices (see Figure 1).

**Figure 1.** Car-Share Parking in Sydney.

Governing parking, and in particular assuring the availability of parking, was critical to car sharing's success in Sydney. Car sharing does not significantly challenge regulatory frameworks or political interests. Car share cars are cars, steered by licensed drivers, on existing roads. As such, these are instances of mundane, small-scale decisions and guiding that becomes bedrocks of mobility transition through adaptation rather than overthrow. Equally important for my argument about framing and categories is that car sharing was not simply seen as a mode of transport. Rather it was the car as a material object – especially when stationary – that required space and thus came into the purview of conventional space allocation and development approval processes. The tools for governing this form of smart mobility already existed, and became adaptable to new imperatives.

## PERSONAL MOBILITY DEVICES

My second example is a less common one – personal mobility devices (PMDs). They are emblematic of new means of vehicle propulsion rather than ICT innovations, yet are likely, in some form, to become part of future transit ecosystems and mobility-on-demand. PMDs are motorized or motor-assisted and personalized modes of transport. Neither bikes nor road vehicles, PMDs are battery-powered motorized devices designed to be used by an individual on footpaths or shared user paths (Rose & Richardson, 2010). Common examples include Segway, YikeBike or two-wheeled 'razor' scooters that have motors (Faulks et al., 2013; Iryo et al., 2012). They allow an individual rider to travel short distances quickly without the physical effort required of cycling, and in some their small size makes transfers between transport modes possible. As small, powered, devices they can bridge the 'first and last mile' – between home and transit and/or transit and work locations. They are potentially an important cog in a smart mobility system, and potential reductions in use of fossil fuels and portability that encourages integration with public transport render these devices more sustainable than the automobile.

Yet PMDs are a marginalized or prohibited transport option in many parts of the world (ACT Government, 2012). In Australia, for example, in all states except Queensland it is illegal to ride a PMD on a public road or footpath. The presentism of transport categories is most starkly in evidence here. In Australia, for example, the number of wheels and motorized character of most of these devices places them into the category 'moped' according to Australian Design Rules. However, since the device does not meet the applicable standards for mopeds, it cannot be registered for, or legally used on roads. Likewise, most of these devices are too dissimilar from bicycles (even electric bicycles, which in NSW are defined by the presence of pedals) to be classified as bicycles. Motorization and wheels also rule out classification as a pedestrian. A pedestrian walks, and by implication, cannot be power assisted. This inability to be squeezed into transport categories has consequences because of the strong nexus between transport infrastructure and transport mode. Footpaths are seen as the principle domain of pedestrians, and regulations are made through pedestrian eyes – especially in relation to speed restrictions.

This is paralleled by legitimate worries on the safety implications of higher speed devices on sidewalks. At the same time, roads are constructed as the domain of cars. Again, speed is illustrative, with limitations placed on what can go on a road dependent on minimum speeds. I am not suggesting these regulations on speed are absurd or unnecessary; ensuring the safety of all travellers is one of the pillars of transport governance. I use the example to demonstrate the power, and path dependence, of the categories of transport governance. The way we think about modes and means of travel – especially in relation to technological additions to human capacities – can shut down opportunities.

Analogous to Jensen et al.'s (2017) work in thinking about bicycles in Copenhagen, we see these devices emerge in transport discourses when they are envisioned differently. The first legalized use of Segways in Australia was announced by the Queensland state government in 2013. This was a political decision and took transport regulation by surprise. As reported at the time, transport and road safety experts within the Queensland public service 'recommended against the approval of Segways for anything but use on private properties because of the threat to public safety and potential for conflict on road-related areas' (Ironside, 2013). It was, however, welcomed by tourist groups, with the decision justified as creating opportunities for tourism operation and jobs. The point to draw from this example is that change not only came from outside transport, but also from a different categorization – Segways were a tool of economic development rather than a means of getting around the city. That is, such devices become regulated not because they were seen to be a desirable addition to the myriad of options for city dwellers to leave their cars at home for short trips, but because they suited the economic imperatives of the state of Queensland (see also Chapters 7 and 8 in this volume).

But before Segways could appear on Queensland streets, appropriate regulations needed to be developed. The first step was to broaden the definition beyond the Segway brand to encompass a generically defined but self-balancing device – a step largely driven by national competition policy and fears that Segway was being given favourable treatment. The second step was to draft regulations to govern the use of these devices, and in these both the longevity and adaptability of transport categories can be identified. We see longevity, in that PMDs were given not their own category of vehicle, but a subset of a mobile body (Figure 2). They were

**Figure 2.** Regulating Motorized Use of Pedestrian Environments.

classed as a type of pedestrian: a pedestrian includes a user of a personal mobility device (Queensland Government, 2015).

But there was also adaptability, with new sets of behaviours and responsibilities for pedestrians introduced. There were detailed specifications of what a PMD is, where it can be used and how it can be used. In terms of what, there are specifications on width, weight, power and potential to be speed limited. They can be used on road-related areas, unless specifically prohibited. Rules on use are particularly pertinent. PMDs can be used on pedestrian environments, though in the context of incredibly strict rules on use. Riders must not travel faster than 12 kilometres per hour, must wear a helmet and must neither travel on a road for more than 50 metres nor be aged under 12. In short, regulatory settings were modified to accommodate these new forms of technology, though not without contestation. PMDs were defined as more like pedestrians than cars and as a result were confined to footpaths and not roads, and riders are called users rather than drivers. Although an 'epistemic experimentation' catapulted PMDs into the transport landscape, its reframing in transport policy confined it to the footpath.

## CONCLUSIONS AND IMPLICATIONS

Through this discussion of car sharing and PMDs I suggest that when smart mobility is framed differently, perhaps outside the traditional categories of transport, or through an apprehension of the need to experiment and innovate, different possibilities emerge. My first point is that governance is already adaptable and flexible, transport policy is constantly changing in small-scale ways that will see it well placed to face the challenges of a smart mobility transition. The accommodation of car sharing into planning policies described here translated into recognition in NSW transport policy, confirming the malleability of policy. What appears less malleable is the ability of governance frameworks to grapple with action that is at once human and technological, a hybrid form in which humans and technologies have agency. The ongoing debate about definitions of drivers and responsibilities in driverless cars is just one example here.

Finally, the example of PMDs and the redefinition of existing categories may be of broader significance in terms of what happens when a new technology is introduced to pedestrian environments. PMDs are far from the strangest objects that have appeared, or are expected to appear, on sidewalks in the future, especially as robotics weaves through all domains of transport. This is especially the case with delivery robots. Already, a number of US cities have welcomed these to their sidewalks, while others, and in particular San Francisco, have banned them outright, even though they were under some form of human control. It will be interesting to see how the politics of this plays out, but also through which categories robots presence on sidewalks is regularized.

## ACKNOWLEDGEMENTS

This research was supported by a Macquarie University Enterprise Partnership Grant, the City of Ryde, and the Australian Research Council (DP170103384).

## REFERENCES

Aldred, R., & Tepe, D. (2011). Framing scrappage in Germany and the UK: From climate discourse to recession talk? *Journal of Transport Geography*, *19*(6), 1563–1569. doi:10.1016/j.jtrangeo.2011.04.008

Australian Capital Territory (ACT) Government. (2012). *Segway review: A review of Segway use and commercialisation in the Australian Capital Territory*. Canberra: Justice and Community Safety Directorate.

Belk, R. (2014). You are what you can access: Sharing and collaborative consumption online. *Journal of Business Research*, *67*(8), 1595–1600. doi:10.1016/j.jbusres.2013.10.001

Blomley, N. (2014). Disentangling law: The practice of bracketing. *Annual Review of Law and Social Science*, *10*(1), 133–148. doi:10.1146/annurev-lawsocsci-110413-030719

Bonham, J., & Cox, P. (2010). The disruptive traveller?: A Foucauldian analysis of cycleways. *Road & Transport Research: A Journal of Australian and New Zealand Research and Practice*, *19*(2), 42–53.

Brown, J. R., Morris, E. A., & Taylor, B. D. (2009). Planning for cars in cities: Planners, engineers, and freeways in the 20th century. *Journal of the American Planning Association*, *75*(2), 161–177. doi:10.1080/01944360802640016

Bulkeley, H., Powells, G., & Bell, S. (2016). Smart grids and the constitution of solar electricity conduct. *Environment and Planning A*, *48*(1), 7–23. doi:10.1177/0308518X15596748

Cohen, M. J. (2012). The future of automobile society: A socio-technical transitions perspective. *Technology Analysis & Strategic Management*, *24*(4), 377–390. doi:10.1080/09537325.2012.663962

Curtis, C., & Low, N. (2012). *Institutional barriers to sustainable transport*. Farnham, Surrey: Ashgate.

Dowling, R., & Kent, J. (2015). Practice and public–private partnerships in sustainable transport governance: The case of car sharing in Sydney, Australia. *Transport Policy*, *40*, 58–64. doi:10.1016/j.tranpol.2015.02.007

Dowling, R., & Simpson, C. (2013). 'Shift – the way you move': Reconstituting automobility. *Continuum*, *27*(3), 421–433. doi:10.1080/10304312.2013.772111

Driscoll, P. A., Theodórsdóttir, Á. H., Richardson, T., & Mguni, P. (2012). Is the future of mobility electric? Learning from contested storylines of sustainable mobility in Iceland. *European Planning Studies*, *20*(4), 627–639. doi:10.1080/09654313.2012.665036

Dudley, G., Banister, D., & Schwanen, T. (2017). The rise of Uber and regulating the disruptive innovator. *The Political Quarterly*, *88*(3), 492–499. doi:10.1111/1467-923X.12373

Faulks, I., Irwin, J., Howitt, R., & Dowling, R. (2013). Electric unicycles, minifarthings and the future of urban transport. *The Conversation*. Retrieved from http://theconversation.com/electric-unicycles-minifarthings-and-the-future-of-urban-transport-13331

Gurran, N., & Phibbs, P. (2017). When tourists move in: How should urban planners respond to Airbnb? *Journal of the American Planning Association*, *83*(1), 80–92. doi:10.1080/01944363.2016.1249011

Haarstad, H. (2016). Where are urban energy transitions governed? Conceptualizing the complex governance arrangements for low-carbon mobility in Europe. *Cities*, *54*, 4–10. doi:10.1016/j.cities.2015.10.013

Henderson, J. (2009). The spaces of parking: Mapping the politics of mobility in San Francisco. *Antipode*, *41*(1), 70–91. doi:10.1111/j.1467-8330.2008.00657.x

Ironside, R. (2013). Transport and main roads staff raised fears about Segways on paths and bikeways with Minister Emerson. The Courier Mail.

Iryo, T., Kusakabe, T., Yamanaka, I., & Asakura, Y. (2012). Effect on travelers' activities and environmental impacts by introducing a next-generation personal transport system in a city. *International Journal of Sustainable Transportation*, *7*(3), 226–237. doi:10.1080/15568318.2013.710124

Jensen, J. S., Cashmore, M., & Elle, M. (2017). Reinventing the bicycle: How calculative practices shape urban environmental governance. *Environmental Politics*, *26*(3), 459–479. doi:10.1080/09644016.2017.1311089

Kent, J., Dowling, R., & Maalsen, S. (2017). Catalysts for transport transitions: Bridging the gap between disruptions and change. *Journal of Transport Geography*, *60*, 200–207. doi:10.1016/j.jtrangeo.2017.03.013

Kent, J. L., & Dowling, R. (2016). 'Over 1000 Cars and No Garage': How urban planning supports car(park) sharing. *Urban Policy and Research*, *34*(3), 256–268. doi:10.1080/08111146.2015.1077806

Lawhon, M., & Murphy, J. T. (2011). Socio-technical regimes and sustainability transitions: Insights from political ecology. *Progress in Human Geography*, *36*(3), 354–378.

McGuirk, P., Dowling, R., Brennan, C., & Bulkeley, H. (2015). Urban carbon governance experiments: The role of Australian local governments. *Geographical Research*, *53*(1), 39–52. doi:10.1111/1745-5871.12098

Potter, S., & Enoch, M. (2016). Paratransit: The need for a regulatory revolution in the light of institutional inertia. *Paratransit: Shaping the Flexible Transport Future* (Vol. 8, pp. 15–34). Bingley: Emerald Group Publishing Limited.

Queensland (QLD) Government. (2015). Rules for personal mobility devices. Retrieved from https://www.qld.gov.au/transport/safety/rules/wheeled-devices/personal-mobility-devices.

Rose, G., & Richardson, M. (2010). Operational impacts of alternative private passenger vehicles. Retrieved from https://www.onlinepublications.austroads.com.au/items/AP-R351-10

Shoup, D. (2014). The high cost of minimum parking requirements. *Parking Issues and Policies* (Vol. 5, pp. 87–113). Bingley: Emerald Group Publishing Limited.

Shove, E., Watson, M., Spurling, N., & Skillington, T. (2015). Conceptualizing connections. *European Journal of Social Theory*, *18*(3), 274–287. doi:10.1177/1368431015579964

Tranter, K. (2017). Disrupting Technology Disrupting Law. *Law, Culture and the Humanities*, 1–14. doi:10.1177/1743872117704925

# CHAPTER 5

# GOVERNING THE RACE TO AUTOMATION

Debbie Hopkins and Tim Schwanen

ABSTRACT

*Automated vehicle technologies dominate many visions of future systems of smart mobility. This chapter uses the approach of Transition Management to explore the multi-actor governance processes around automated vehicle technologies in the United Kingdom (UK), with specific attention being paid to the role of the UK government. It shows the relatively comprehensive approach to automated vehicle innovation that has been adopted by the UK government, emerging across multiple domains including the creation of positive discourses around automation, and the facilitation of network building and demonstration projects. Framed by the Transition Management cycle of strategic, tactical, operational and reflexive activities, the chapter argues for greater integration across the levels of the cycle, and experimentation that moves beyond technological capability, to include the heterogeneous publics, in a more diverse set of roles than the current framing of 'potential technology adopter'. The chapter points to the techno-optimism displayed by governments participating in the international race to vehicle automation, often with dual roles as both producers and consumers, and suggest that greater inclusivity, democracy, diversity and openness in the innovation process may contribute to context sensitive implementation.*

**Keywords:** Automation; Transition Management; automated vehicles; automobility

## INTRODUCTION

Automated vehicles have become central to visions of future mobility. It has been proposed that the transition to automated mobility will be the most significant change to the system of mobility since the internal combustion engine enabled motorized mobility. Advocates of automated vehicles (also known as 'driverless', 'self-driving' and 'autonomous' vehicles) claim that these innovations will revolutionize the way people and goods move within and between towns and cities. Claims abound of environmental, social and economic benefits arising from these innovations contributing to greater quality of life, and thus highlighting their potential to transition mobility towards sustainability in its broadest sense. Nonetheless, the emergence of automated vehicles is situated within the system of automobility (Urry, 2004) which prioritizes and rewards private, motorized transport. And while automated vehicles have been framed as disruptive technologies by some (e.g., Arbib & Seba, 2017), others have argued that smart and automated innovations may further cement the longevity of automobility (Schwanen, 2016) and thereby reinforce many of its negative externalities. In addition to questions of disruptiveness, the race to automation has brought with it a range of questions about the suitability of traditional systems of governance.

As (sustainability) transitions are systemic and by definition complex, uncertain and contested, traditional governance processes are poorly suited. Reflexive governance has been identified as one way through which complex issues can be understood, integrating a range of perspectives, expectations and strategies (Voß & Bornemann, 2011). Different designs of reflexive governance have emerged. Transition Management (TM) analytically frames transitions as structural changes in complex, societal systems. TM suggests that by better understanding 'the origin, nature and dynamics' of transition processes, actors will be able to both anticipate and adapt with potential to influence the speed and direction of transition (Loorbach, Frantzeskaki & Huffenreuter, 2015). In this chapter, we suggest that TM's concepts and ideas can be used as a heuristic framework with which to interpret and evaluate currently unfolding governance processes in automated mobility. While the ability of automated technologies to contribute to a *sustainability* transition is, as yet, unclear, the transformation towards automation can still fruitfully be analysed using TM because of the complexities and uncertainties involved in the emergence and potential diffusion of the technology within the socio-technical system of mobility.

## TRANSITION MANAGEMENT, REFLEXIVE GOVERNANCE AND EXPERIMENTATION

Both top-down steering by government, i.e., social change through government policy, and the free-market approach, i.e., social change by way of market forces, are broadly ineffective in producing sustainable societies (Loorbach, 2010), yet the importance of both in the governance of societal change has been documented (Meadowcroft, 2005). Thus, new modes of governance are required that balance state, market and society (Loorbach, 2010). While articulated goals, knowledge and power differ between approaches, it is broadly agreed that participation, experimentation and collective learning need to be central features of governance processes (Voß & Bornemann, 2011). From this perspective, societal change is seen to result from a range of different forces, and reflexive governance approaches seek to embrace the heterogeneity and multi-dimensionality of problems, and the range of possible futures.

TM was introduced as a new governance approach for sustainable development (Loorbach, 2010) and as a form of reflexive governance. It is a governance framework that aids instrument choice (Kemp & Loorbach, 2003), and shapes innovation processes and structural transitions within socio-technical systems, such as mobility (Kemp, Loorbach & Rotmans, 2007). TM is not, however, focused on achieving predefined outcomes through processes of planning and control but rather on goal-oriented modulation (Kemp & Loorbach, 2006). It focuses on influencing 'persistent societal problems' (Rotmans & Loorbach, 2008) and asks fundamental questions of whether transitions can be managed. Emerging from complexity theory, management in this sense speaks of influencing processes of change, from one state to another (Rotmans & Loorbach, 2008). While some have identified TM as 'incremental politics', others have situated it 'in the middle ground between planning and incrementalism' (Rotmans, Loorbach & Kemp, 2007; Voß & Bornemann, 2011, p. 8).

Transitions are not the result of a single change (e.g., policy or technology) but emerge from developments across domains including, but not limited to, technologies, the economy, user practices, cultural values and meanings, and knowledge. The TM cycle (Figure 1) translates the basic principles of the management of transitions into four operational governance activities that exhibit distinct characteristics, including types of actors involved and levels at which they operate (Loorbach & Rotmans, 2010):

1. *Strategic*: societal system levels, with long-time horizons relating to problem structuring, activities include the establishment of a transition arena and the construction of visions of alternative futures;

**Figure 1.** The Transition Management Cycle.

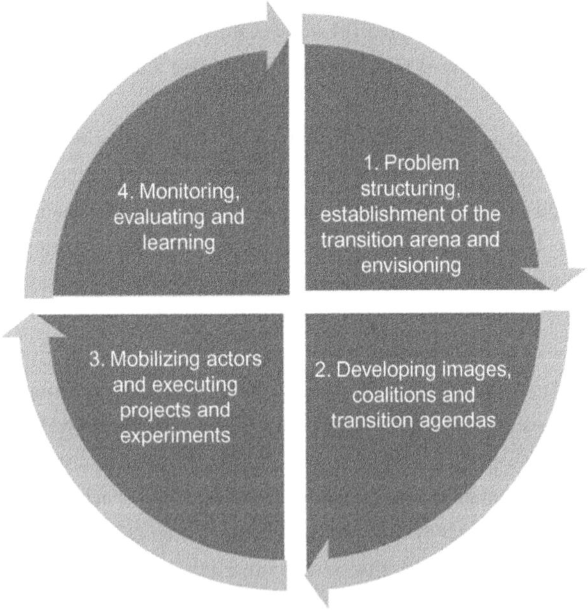


*Source*: Adapted from Loorbach and Rotmans (2010, p. 238).

2. *Tactical*: sub-system-level activities which construct or destruct system structures, activities include the development of coalitions, transition agendas and images;
3. *Operational*: short-term and everyday decisions and actions, including the mobilization of actors and the execution of projects and experimentation and
4. *Reflexive*: the use of debate, research and assessment to evaluate the various levels and their interplay by way of monitoring, evaluation and learning.

While these stages are often depicted as a sequence of events, it is the need for connection between the events rather than the order in which they are completed that is important. TM plays out over the different levels and time scales, with a focus on long-term processes of learning. From experimental applications of TM, a number of guiding principles have emerged, including inclusivity, democracy, diversity and openness, speed of change, social learning and capacity building, the resilience of the regime, context-specific learning and empowerment of frontrunners (Loorbach et al., 2015). These principles can help to develop context-sensitive implementation across different settings (e.g., urban to rural, local to national, energy, mobility, education and welfare).

Experimentation is a critical feature of TM, particularly in the pre-development phase of transition (Kemp & Loorbach, 2003). Transition experiments require both time and resources to establish within niches before becoming embedded in existing regimes. Strategies to achieve this consist of learning from the experiment (deepening), repeating based on earlier learning, experimenting in a different context (broadening) and applying the successful experiment at a larger scale (scaling up). Transition experiments are taken to be 'high-risk' with high chances of failure, but social learning objectives can contribute to systemic sustainability goals and transitional pathways (Loorbach et al., 2015).

Demonstration projects have become an important part of the experimentation process, and can be practiced in different ways, for instance, by *showing* a (near) completed artefact to a particular audience, or as a stage in the developmental process. TM suggests that a multiplicity of experiments form a portfolio that works to reinforce one another (Loorbach et al., 2015), and develop collective and shared learnings, thereby contributing to the predetermined sustainability objectives in detectable and measurable ways. Local experimentation, it is argued, needs to inform national policy and system innovation, with coordination amongst policies across scales (Kemp & Loorbach, 2003).

## AUTOMATING TRANSPORT IN THE UNITED KINGDOM

The future of transport has been a technological and socio-cultural fascination for at least the last century. This fascination has been fuelled by science fiction imaginaries from popular entertainment (e.g., *The Jetsons*, *The Fifth Element*), with visions including robotic, automated and autonomous systems in different forms and with a range of capabilities. Testing and experimentation of automated vehicles has been occurring in the UK, albeit intermittently, since the 1960s, when government-funded projects trialled 'drive-by-wire' technologies that enabled driverless, fully automatic vehicles (Reynolds, 2001). Trials and demonstrations were conducted on both private and public roads including a section of the M4 motorway, with the hardware embedded into the infrastructure as the M4 was constructed. The experiments adapted existing vehicle models – a Citroen DS19 saloon, a Standard Vanguard estate and an Austin Mini – with different automated technologies. The research by the Crowthorne Laboratory, at the time a government agency, was predicated on anticipated motorway congestion, and part of a suite of experiments designed to pre-empt contemporary transport problems. Their research suggested that automated technologies could increase road capacity by 50%, and prevent 40% of accidents. However, funding for the research was withdrawn in the mid-1970s when, in the context of economic recession, the technologies were deemed 'unnecessary and unaffordable' (Reynolds, 2001).

More recently, advances in deep machine learning by ICT providers and software companies have stimulated a new wave of interest in so-called 'driverless' technologies. Automated innovations have emerged from incumbent car manufacturers (e.g., Volvo), new actors (e.g., Tesla) and partnerships between incumbents and new entrants. Examples include Daimler partnering with Uber, General Motors purchasing the start-up 'Cruise Automation' in 2016 and Ford announcing a USD 1 billion investment in 'Argo AI' (Murphy, 2017). These partnerships respond to the need for hybrid expertise of traditional vehicle manufacturing processes, economies of scale, and new demands for ICT know-how, code and algorithms. Incumbent vehicle manufacturers, including BMW, Volkswagen and Nissan, have also been working on automated vehicles themselves for more than a decade. These efforts have been accelerated and extended in recent years under pressure from new entrants such as Waymo and Uber.

The rise in public consciousness and subsequent hype about the capabilities of automated vehicles has been coupled with real-world trials and attention from the mass media. Robotics, Artificial Intelligence and automation are also envisaged by the UK government as a centrepiece of the post-Brexit economy (HM Government, 2017). The focus on automation including research and development, manufacturing and adoption can also be seen as a response to the post-Global Financial Crisis (GFC) economy, domestic regional inequality and the dismantling of indigenous vehicle manufacturing during the 1980s (Harvey, 2005). Consequently, the government has been actively involved in a range of activities, including:

- future visioning and the creation of positive discourses around automation;
- facilitation of network building between incumbents, start-ups and other actors;
- creation or amendment of domestic legislation (e.g., licensing and insurance);
- development of a Code of Practice for testing of automated vehicle technologies;
- engagement with international agencies to create or amend international standards and/or legislation and
- technology trials and demonstration projects.

In short, the UK government has adopted a relatively comprehensive approach to automated vehicle innovation, with activities emerging in multiple domains. The merits of this approach can be reflected on more critically using TM as heuristic framework.

## TRANSITION MANAGEMENT AND THE RACE TO THE FUTURE

We will now adopt the four stages of the TM cycle: strategic, tactical, operational and reflexive, depicted in Figure 1 to structure our evaluation of the race to automation.

### Strategic

At a strategic level, the government's focus on problem definition, the establishment of a transition arena and vision building have been articulated in key publications (e.g., HM Government, 2015a, 2017). These documents communicate the government's position in harnessing benefits from automation technologies as both producers and consumers, and the various roles the government has taken, and intends to take to this end. In 2015, the government released two reports outlining the 'Pathway to Driverless Cars' (HM Government, 2015a, 2015b). These documents serve as action plans with a driverless future predetermined: one report is a more general summary characterizing the technology, and identifying challenges and the opportunities to development and adoption (HM Government, 2015a), while the other is a detailed report examining regulatory frameworks (HM Government, 2015b). Both reports are focused on overcoming barriers and situating the UK as a centre for technology development, testing and demonstration, largely achieved by way of comparison with other regions, countries and states actively competing in the global race (e.g., the United States – particularly the states of Nevada and California, Germany, Sweden, Japan and South Korea).

The Industrial Strategy Green Paper (HM Government, 2017) consulted not only on ten pillars by which the modern industrial strategy can 'increase productivity and drive growth across the whole country', but also on the technologies which could be supported by way of the Industrial Strategy Challenge Fund. In particular, the Industrial Strategy Green Paper views automated vehicles as a way through which the *productivity gap* can be tackled, and as a mechanism to drive growth. These documents make explicit the underlying motivations of the UK government relating to the development of niches with less importance given to the claims of contributing to environmental or social sustainability. Thus, the UK government's response has been centred on a producer logic of creating conditions for technology emergence and diffusion, rather than system-wide, long-term visioning, and a critical analysis of the role automation may play in future mobility.[1]

## Tactical

The tactical level focuses on processes of agenda and coalition building, which occur at the sub-system level. This includes negotiations between key stakeholders and actors. Activities at the tactical level often include transformations of existing structures and institutions in line with articulated strategic goals. The development of networks and coalitions is important to activities at this level as these linkages can provide mechanisms for change. Innovate UK and the Transport System Catapult[2] have helped to bring together key industry actors, but their remit, and subsequent coalition building efforts appear to focus on expert dialogue and the harnessing of established capabilities to compete in the global race to automation. Processes of agenda building, negotiating, networking and coalition building are also the concern of the Centre on Connected and Autonomous Vehicles (CCAV), part of the Department for Transport and the Department for Business, Energy and Industrial Strategy, which was established in 2015 to work across government to aid early market entry for automated vehicle technologies. This partnership of government departments somewhat tellingly points to the intention of the UK government to be active in the business development and innovation of automated vehicle technologies. Indeed, the remit of the Centre is to 'help ensure that the UK *remains* a world leader in developing and testing connected and autonomous vehicles' (HM Government, 2015a, emphasis added). Thus while new coalitions, partnerships and networks may be enabled by the tactical processes of TM, the selectivity in coalition building evidenced here may work against TM principles of inclusivity, democracy, diversity and openness (Loorbach et al., 2015).

## Operational

Experimentation is central to the TM literature, and a key part of the race to automation, evidenced in the UK and elsewhere not only by industry actors but also governmental approaches to automated vehicle development. Public investment in demonstration projects and field trials has been noted in other innovation pathways including renewable energy sources, with government funding 'in a little understood "black box" process to effect technology improvements and change the complex industry, market and regulatory barriers' to new technologies (Harborne & Hendry, 2009, p. 3581). The operational phase of TM, through the mobilization of actors, and the execution of projects and experimentation is underpinned by the funding rounds provided by the UK government. These have worked to develop partnerships across industry, research and government, but the degree to which learning feeds into, or is informed by, other stages of the TM cycle is not evident.

The UK government has provided funding to stimulate domestic innovation, including driverless vehicle trials, and three Connected and Autonomous Vehicle (CAV) competitions (2016, 2017, 2018) offering matched industry funding. In 2014, Innovate UK, the UK's innovation agency, led a competition for £19 million of funding, and identified four sites of demonstration constituting partnerships between private companies, consultancies, universities and local authorities, and culminating in three trials lasting between 18 and 36 months. Bristol, the Royal Borough of Greenwich (south-east London), Coventry and Milton Keynes were selected as sites for learning and experimentation at the 'frontline of new economic, cultural, political and societal configurations in cities' (Karvonen & Van Heur, 2014, p. 389) – for automated vehicle innovations and with the intention of establishing 'the UK as the global hub for the research, development and integration of driverless vehicles and associated technologies' (Innovate UK, 2014).

The funding for demonstration projects was recognized by the then – Business Secretary Sajid Javid as a '*further sign of our commitment to making sure we're creating opportunities for UK businesses to thrive and attract global investment in world-class technology*' (Innovate UK, 2014), pointing to the economic motivations underlying the demonstrations, and their role in the international race. In addition to the government-led trials, other industry-led demonstrations are underway, for instance from technology-developing start-ups, and incumbent manufacturers, including Volvo's 'Drive Me London' programme (Table 1). The use of geographically dispersed and diverse demonstrations fits with the TM philosophy.

**Table 1.** Public Demonstration Projects and Trials of Automated Passenger Vehicles.

| Name | Time Frame | Places and Sites | Technologies | Actors | Government Funding |
|---|---|---|---|---|---|
| LUTZ Pathfinder Project | 2014–2016 | Milton Keynes, Pavements | Passenger pods | Oxbotica, Oxford Robotics Institute, Transport System Catapult, Milton Keynes Council | Transport System Catapult |
| Venturer consortium | 2016–2018 | Bristol and South Gloucestershire, University Campus, Urban public roads | Connected and autonomous passenger vehicles | University of Bristol, Bristol City Council, AXA, First Group, University of the West of England, Atkins, Fusion Processing, South Gloucestershire Council, Williams, BAE Systems, Bristol Robotics Group | UK Driverless Car Trial Competition |
| Autodrive consortium | 2016–2018 | Nuneaton, Coventry and Milton Keynes | Driverless passenger pods and adapted private passenger vehicles (Ford, Jaguar Land Rover and Tata Motors) | ARUP, AXA, Coventry City Council, Ford, Gowling WLG, MIRA, Jaguar Land Rover, Milton Keynes Council, RDM Group, TATA Motors, Thales, The Open University, Transport Systems Catapult, University of Cambridge, University of Oxford | UK Driverless Car Trial Competition |

**Table 1.** *(Continued)*

| Name | Time Frame | Places and Sites | Technologies | Actors | Government Funding |
|---|---|---|---|---|---|
| GATEway consortium | 2016–2017 | Shared environment with pedestrians and cyclists, Greenwich London and Heathrow | Passenger and freight: Passenger pods, public driverless shuttles, autonomous-enabled Toyota | TRL, Royal Borough of Greenwich, Royal College of Art, Greenwich University, RSA, O2, Oxbotica, Imperial College London, Heathrow Airport, GoBotix, Commonplace, Westfield, Shell | UK Driverless Car Trial Competition |
| MOVE_UK consortium | 2016–2019 | Greenwich, public roads | Passenger vehicles with Automated Driving Systems (ADS) | Bosch, Jaguar Land Rover, Direct Line Group, The Floow, the Royal Borough of Greenwich, TRL | Intelligent Mobility Fund[a] |
| Volvo Drive Me London | 2017–2018 | West London public roads | Adapted Volvo XC90 (4x4) | Volvo | Not clear |
| Streetwise consortium | 2017–2019 | London | Autonomous and connected vehicles (SAE 'level 4'[b]) | FiveAI, Direct Line, the University of Oxford, Transport for London and the Transport Research Laboratory | CAV2 |

| | | | | | |
|---|---|---|---|---|---|
| DRIVEN consortium | 2017–2019 | Oxford to London, RACE – private testing space, and motorway | Autonomous and connected vehicles (SAE 'level 4') | Oxbotica, the Oxford Robotics Institute, XL Catlin, Nominet, Telefonica, the Transport Research Laboratory, UKAEAs RACE facility, Oxfordshire County Council, Transport for London and Westbourne Communications | Innovate UK & CCAV |
| Insight consortium | 2017–2019 | Birmingham, City centre environment | Autonomous shuttles | Birmingham City University and Westfield Sportscars Ltd, Heathrow Enterprises Ltd, Fusion Processing Ltd, Creative Example Ltd and Conigital Ltd | Intelligent Mobility Fund |

[a]In 2016, eight projects were awarded £20 million of funding to develop automated vehicle technologies through the Intelligent Mobility Fund. This included some public demonstrations and technology trials (e.g., MOVE_UK) and mobile platforms (e.g., i-MOTORS).
[b]The Society of Automotive Engineer (2014) levels of automation, ranging from zero to five, depict the varying degrees of motor vehicle automation in relation to human control of driving functions.

However, while multiple technologies are tested in experiments involving different actor constellations, it is notable that to date, all government-funded demonstrations are located in southern England, which may reinforce the asymmetric economies of the south of England and the remainder of the United Kingdom.

Places of experimentation, and the role of users, present two concerns at the operational level. There has been a focus to date on *urban* experimentation of automated technologies; however, given the complexities of urban transport systems, urban traffic may well be the most difficult to automate (Schwanen, 2017). Perhaps in response to this, there have been steps towards experimentation on motorways for both passenger vehicles (e.g., DRIVEN consortium) and freight vehicle platooning.

A clear absence in experimentation is the active participation of the public, who are effectively configured as 'future technology adopters'. As the trials have progressed through time, there has been more recognition of users as part of the innovation pathway, and thus the potential for a broadening of learning and opportunities for greater inclusivity. For instance, the Insight consortium research focuses on accessibility for independent travel, and other user-centred projects funded through the same 'Intelligent Mobility Fund' mechanism (e.g., Flourish, http://flourishmobility.com/) are highlighting the need to better understand the social equity dimensions of automated vehicle innovations. However, there remains a clear prioritization of experts from government, industry and academia in the discourse relating to robotics and automation and little public engagement beyond public perception surveys (Kyriakidis, Happee, & de Winter, 2015; Schoettle & Sivak, 2014), consultation on the Industrial Strategy Green Paper (HM Government, 2017), and exposure to the technologies through public trials. As previously reported with electric vehicles (Bergman, Schwanen, & Sovacool, 2017), a wider range of roles, including as citizens, is not currently envisaged in automated vehicle experimentation.

### Reflexive

The *reflexive* stage in the cycle is not yet well articulated, with learning appearing to relate to international government policy and regulation, and industry developments. For instance, the UK government Pathway to Driverless Cars regulatory review (HM Government, 2015b) provides a detailed examination of international government regulations to driverless vehicle technologies, to aid the development of domestic policy. This provides some evidence of the competitive stance taken by the government with relation to gaining domestic innovation, research and development, and demonstration projects. It appears that discourses of international competition and economic significance, and the frame of a race towards automation have created a sense of urgency that results in a privileging of implementation of legislation and cutting of 'red tape' over public participation in the innovation trajectory.

So-called 'development rounds' are important steps in the TM process involving monitoring and evaluation of the experimentation (Kemp & Loorbach, 2003). Part of the learning achieved in this process is an evaluation of whether the transition objectives have been achieved. This critical and self-reflexive process offers opportunities for restructuring the problem framing. The diversity of private-sector actors, and the overlap between demonstrations in terms of technologies (e.g., vehicles, sensors) used, and consortium partners (Table 1), may suggest learnings between projects and the collaborative development of technologies. Insofar as lessons have been learnt, these are not widely circulated and not in the public domain, no doubt in part because of commercial sensitivities and the perceived importance to not advance the position of perceived competitors. Moreover, as far as automated vehicle developments in the UK are concerned, there appears to be little space for critical interventions, big picture thinking, reflexivity or indeed failure. This can be attributed at least in part to the way in which the smart mobility transition is framed in the context of neo-liberal ideals and dependence on private-sector interests.

## Interconnections

The interconnections between the four TM levels are important to long-term learning and other TM outcomes. There are multi-level dynamics, whereby strategic activities can influence tactical, operational, and reflexive activities through top-down processes, yet experimentation and learning from the operational and reflexive levels can also put pressure on the strategic level by way of bottom-up forces. With automated vehicles in the UK, the TM stages do not appear to be well aligned and integrated. It is not immediately clear how the monitoring and evaluation of the demonstration projects will contribute to wider learnings and reflexive governance of the transition to automated vehicles. These insights point to the need for multi-level coordination with local experimentation informing the development of national policy, and coordination across scales, including with international policies as 'go-it-alone policies' can be harmful unless there are clear first-mover advantages (Kemp & Loorbach, 2003).

## CONCLUSIONS

The frame of a *race to automation* enforces the need to participate, or risk being left behind. The innovation race can reinforce techno-optimism, particularly in national economies, such as the UK, that remain scarred by the 2007–2008 financial crisis. At present, the diversity of actors and demonstration projects in the UK offer opportunities for learning, albeit within the selective coalitions. The current transition experiments are focused on 'deepening' learning from demonstrations and trials, with a need for 'broadening' and 'scaling up' the experimentation. This should move beyond technological capability, and frame public input in a more diverse set of roles (beyond 'adopter'). Nevertheless, notions of inclusivity and democracy, which are guiding principles of TM, are conspicuously absent from the current approach to automated vehicle innovation in the UK and need to underpin experiments as they occur in different contexts and at wider scales.

At a minimum, democratic discussions with due consideration for potential winners and losers across spatial and temporal scales are required at an early stage of development. There is an apparent 'post-political' consensus (Mouffe, 2005; Swyngedouw, 2010) about the centrality of markets and expert judgement that forecloses radical dissent (Schwanen, 2017), and which seems set to steer the development of automated futures in ways which the TM literature suggests may be sub-optimal. Further consideration of the (post-) political and TM is required to underscore the political dimensions of automated vehicle transition dynamics. Moreover, Robust interconnections between activities across the four levels (strategic, tactical, operational and reflexive) could enhance or accelerate a transition to automated mobility. Nevertheless, critical questions about whether an automated mobility future can respond to the environmental, social and economic challenges created and reinforced by automobility remain unanswered.

## NOTES

1. The House of Lords Science and Technology Select Committee (2017) conducted a thorough assessment review of connected and autonomous vehicles (CAVs); however, it is unclear the impact this has had or will have on public automated vehicle discourse.

2. The Transport Systems Catapult is a not-for-profit organization overseen by Innovate UK, with a vision of 'global leadership in Intelligent Mobility, promoting sustained economic growth and wellbeing through integrated, efficient and sustainable transport systems'.

## ACKNOWLEDGEMENTS

This work was supported by the Research Councils United Kingdom (RCUK) Energy Programme (grant EP/K011790/1, 'Centre on Innovation and Energy Demand' (CIED)).

## REFERENCES

Arbib, J. & Seba, T. (2017). *Rethinking transportation 2020–2030: The disruption of transportation and the collapse of the internal-combustion vehicle and oil industries*. RethinkX Disruption Report. May 2017. Retrieved from https://static1.squarespace.com/static/585c3439be65942f022bbf9b/t/591a2e4be6f2e1c13df930c5/1494888038959/RethinkX+Report_051517.pdf?pdf=RethinkingTransportation

Bergman, N., Schwanen, T. & Sovacool, B. (2017). Imagined people, behaviour and future mobility: Insights from visions of electric vehicles and car clubs in the United Kingdom. *Transport Policy*, *59*, 165–173.

Harborne, P. & Hendry, C. (2009). Pathways to commercial wind power in the US, Europe and Japan: The role of demonstration projects and field trials in the innovation process. *Energy Policy*, *37*(9), 3580–3595.

Harvey, D. (2005). *A brief history of neoliberalism*. Oxford: Oxford University Press.

HM Government. (2015a). *Driverless vehicles: Connected and autonomous technologies*. London: Department for Transport. 14 August 2015. Retrieved from: https://www.gov.uk/government/collections/driverless-vehicles-connected-and-autonomous-technologies

HM Government. (2015b). *The pathway to driverless cars: A detailed review of regulations for automated vehicle technologies*. London, UK: Department for Transport. Retrieved from: https://www.gov.uk/government/uploads/system/uploads/attachment_data/file/401565/pathway-driverless-cars-main.pdf

HM Government. (2017). *Building our industrial strategy*. Green Paper. January 2017. London: HM Government. Retrieved from https://beisgovuk.citizenspace.com/strategy/industrial-strategy/supporting_documents/buildingourindustrialstrategygreenpaper.pdf

House of Lords Science and Technology Select Committee. (2017). *Connected and autonomous vehicles: The future?* 2nd Report of Sessions 2016-17. 15 March 2017. Authority of the House of Lords. London: HM Government. Retrieved from https://publications.parliament.uk/pa/ld201617/ldselect/ldsctech/115/115.pdf

Innovate UK. (2014). *Driverless cars: 4 cities get green light for everyday trials*. 3 December 2014. London: HM Government. Retrieved from https://www.gov.uk/government/news/driverless-cars-4-cities-get-green-light-for-everyday-trials

Karvonen, A. & Van Heur, B. (2014). Urban laboratories: Experiments in reworking cities. *International Journal of Urban and Regional Research*, *38*(2), 379–392.

Kemp, R. & Loorbach, D. (2003). *Governance for sustainability through transitions management*. Paper for EAEPE 2003 Conference, Maastricht, the Netherlands, 7–10 November 2003.

Kemp, R. & Loorbach, D. (2006). Transition management: A reflexive governance approach. In J.-P. Voß, D. Bauknecht, & R. Kemp (Eds.), *Reflexive governance for sustainable development*. Cheltenham: Edward Elgar.

Kemp, R., Loorbach, D. & Rotmans, J. (2007). Transitions management as a model for managing processes of co-evolution towards sustainable development. *International Journal of Sustainable Development and World Ecology*, *14*(1), 78–91.

Kyriakidis, M., Happee, R. & de Winter, J. C. F. (2015). Public opinion on automated driving: Results of an international questionnaire among 5000 respondents. *Transportation Research Part F: Traffic Psychology and Behaviour*, *32*, 127–140.

Loorbach, D. (2010). Transition management for sustainable development: A prescriptive, complexity-based governance framework. *Governance: An International Journal of Policy, Administration and Institutions*, *23*(1), 161–183.

Loorbach, D., Frantzeskaki, N. & Huffenreuter, R. L. (2015). Transition management: Taking stock from governance experimentation. *The Journal of Corporate Citizenship*, *58*, 48–66.

Loorbach, D. & Rotmans, J. (2010). The practice of transition management: Examples and lessons from four distinct cases. *Futures*, *42*, 237–246.

Meadowcroft, J. (2005). Environmental political economy, technological transitions and the state. *New Political Economy*, *10*(4), 479–498.

Mouffe, C. (2005). *On the political*. Abingdon: Routledge.

Murphy, M. (2017). Ford knows where to invest to win the self-driving car race – and it's not in cars. *Quartz*, 11 February. Retrieved from https://qz.com/908151/ford-knows-where-to-invest-to-win-the-self-driving-car-race-and-its-not-in-cars/

Reynolds, J. (2001). Cruising into the future. *The Telegraph*, 26 May. Retrieved from http://www.telegraph.co.uk/motoring/4750544/Cruising-into-the-future.html

Rotmans, J. & Loorbach, D. (2008). Transition management: Reflexive governance of societal complexity through searching, learning and experimenting. In J. C. J. M. Van den Bergh & F. R. Bruinsma (Eds.), *Managing the transition to renewable energy: Theory and practice from local, regional and macro perspectives*. Cheltenham: Edward Elgar.

Rotmans, J., Loorbach, D. & Kemp, R. (2007). *Transition management: Its origin, evolution and critique*. Workshop on Politics and Governance in Sustainable Socio-Technical Transitions, Berlin, Germany, 19–21 September 2007.

Schoettle, B. & Sivak, M. (2014). A survey of public opinion about autonomous and self-driving vehicles in the US, the UK and Australia. UMTRI-2014-21, July. Retrieved from https://deepblue.lib.umich.edu/bitstream/handle/2027.42/108384/103024.pdf?sequence=1&isAllowed=y

Schwanen, T. (2016). Rethinking resilience as capacity to ensure; City: Analysis of urban trends. *Culture Theory, Policy, Action*, *20*(1), 152–160.

Schwanen, T. (2017). Peak car? Driverless technology may actually accelerate car ownership. *The Guardian*, Tuesday 18 July. Retrieved from https://www.theguardian.com/sustainable-business/2017/jul/18/peak-car-driverless-technology-car-ownership-tesla-nissan-uber

Society of Automotive Engineers [SAE]. (2014). Automated driving. Retrieved from https://www.sae.org/misc/pdfs/automated_driving.pdf

Swyngedouw, E. (2010). Apocalypse forever? Post-political populism and the spectre of climate change. *Theory, Culture and Society*, *27*(2–3), 213–232.

Urry, J. (2004). The 'system' of automobility. *Theory, Culture & Society*, *21*(4–5), 25–39.

Voß, J.-P. & Bornemann, B. (2011). The politics of reflexive governance: Challenges for designing adaptive management and transition management. *Ecology and Society*, *16*(2), 1–24.

# CHAPTER 6

# WHO BENEFITS FROM SMART MOBILITY POLICIES? THE SOCIAL CONSTRUCTION OF WINNERS AND LOSERS IN THE CONNECTED BIKES PROJECTS IN THE NETHERLANDS

Edgar Salas Gironés and Darja Vrščaj

ABSTRACT

*We analysed the unequal treatment of target groups and the role of technology in Dutch smart mobility policy, by looking into the connected bike projects of Maastricht and Brabant. In doing so we combined insights from the Social Construction of Policy Design frameworks and Science and Technology Studies. We identified four target groups, receiving a differential treatment in policy. Difference in treatment is driven by the variety of incentives used to encourage behaviour change. We conclude that car users are the winners, while students benefit the least from these projects. In this preferential treatment, technology plays a crucial role. This raises pertinent questions about social equity and the contribution to sustainability of smart mobility technologies in the mobility system.*

**Keywords:** Connected bike; Social construction and policy design; target populations; cycling policy; science and technology studies

## INTRODUCTION

The current mobility system faces several challenges, including sustainability, efficiency, congestion and safety (Geels, 2012). Thus, sustainability transitions and mobility scholars have promoted a transformation towards a smarter and more sustainable system (Geels et al., 2012). One approach to achieve this transformation is by incorporating smart mobility solutions and behavioural change. In this chapter we study the 'connected bike': a(n) (e-)bike with tracking technology recording and providing feedback on the user's activity. This technology provides multiple benefits to the mobility system, in terms of helping to reduce congestion and improve accessibility; and for users, including health benefits, reduced travel times and wellbeing.

Promoting the adoption of the connected bike relies heavily on the capacity of policymakers to induce behavioural change in target groups. Such selection raises questions about equity and justice relating to these beneficiaries: What target groups are included and excluded by policymakers, and why? How are these groups treated in policy? Is there any preferential treatment among the selected groups? To answer these questions, policy scientists have relied on the Social Construction and Policy Design (henceforth SCPD) framework. This framework indicates that preferential treatment of these groups can be explained by their social construction, referring to their social and cultural characterizations based on values, symbols and images (Pierce et al., 2014). The SCPD framework has been extensively used in other policy domains but its application in the transport domain remains limited (cf. Ingram, Schneider & Deleon, 2007). Previous applications of this framework have also neglected the role of technology in providing differential treatment to target groups. We address this by incorporating insights from Science and Technology Studies (hereafter STS), to understand the role of technology in society (MacKenzie & Wajcman, 1999).

This chapter applies the SCPD and STS methods to analyse the differential treatment of target groups and the role of technology therein. It takes two Dutch connected bike projects from Beter Benutten (BB), a national Dutch programme to reduce traffic congestion and travel time by behavioural change, the 'Burn Fat Not Fuel' (BFNF) project in Maastricht and 'B-riders' in Brabant. This allows for a comparative case study approach as they are both contextually similar (geographically adjacent and localized in the same jurisdiction of the national road authority (Rijkswaterstaat)) but have distinct approaches to behavioural change, decision-making mechanisms and technological choices.

Our research question is 'what is the role of technology in the differential treatment of target groups in the Beter Benutten connected bike projects in the regions of Maastricht and Brabant?' To answer this inquiry, we address the following questions: 'which target groups are identified in these regions?'; 'what type of preferential treatment occurs in the target groups identified?' and 'what is the role of science and technology therein?' To answer these questions, we interviewed project stakeholders and conducted a literature review. In this chapter, we unravel policymaking implications of the smart mobility technologies, particularly regarding questions of equity and justice, and simultaneously reflect on the dilemmas that policymakers may face in the diffusion of these technologies. The structure of the chapter is as follows: the second section outlines the theoretical basis for our research. The third section describes the method. The fourth section introduces the case study and the fifth section contains the findings. The chapter then draws its conclusions in the sixth section.

## COMBINING SCPD AND STS FOR STUDYING SMART MOBILITY TARGET GROUPS

Target groups are 'the persons, groups, or firms selected for behavioural change by public policy initiatives' chosen to receive wins and losses through the multiple elements of the policy design (Schneider & Ingram, 1993, p. 334). Putatively, target group treatment should be based on policy expertise. However, in practice, some target groups receive preferential or negative treatment. To explain this unequal treatment, the SCPD framework looks into how policymakers distribute differentially wins and losses to target groups based on social constructions. By social constructions, we refer to the social and cultural characterizations that policymakers (and society overall) have about social groups, based on attributing values, symbols and images (Pierce et al., 2014). These can also be seen as the stereotypes attributed to beneficiaries of policies. These characteristics are used as a selection-criterion for specific kinds of treatment or for access to policy-regulated resources.

Following the SCPD, two dimensions influence target groups treatment: their social construction in positive or negative terms, and their level of power (Ingram et al., 2007). Classical positive constructions include 'honest', 'socially committed', 'trustworthy' or 'in need of support', while negative characterizations are 'dishonest', 'inconsiderate' or 'irresponsible'. Positively constructed target groups include elderly, entrepreneurs and orphans, while negatively constructed groups encompass thieves and terrorists. The second dimension is their level of power, which is reflected in elements such as group size and wealth, as well as their capacity to mobilize and influence public debate (Ingram & Schneider, 2013). While powerful groups may contest unfavourable treatment in policy, less powerful groups may have limited capacity to do so.

We can explore how this social construction results in differential treatment by following the distribution of wins and losses. By wins, we refer to policy measures aimed at benefiting or rewarding a target group, while by losses we refer to measures resulting in the punishing or burdening a target group. We adopt the SCPD theory's (generally implicit) distinction between materialized and rhetorical wins or losses. Materialized wins and losses are tangible policy measures for targets groups, such as fiscal incentives or taxes. These are best understood in contrast to rhetorical wins, which do not materialize, either because they are promised but never put into practice or are intended to be symbolic. Based on these two dimensions, SCPD policy scholars have proposed four 'ideal types' of target groups (Ingram et al., 2007; Schneider & Ingram, 1993):

1. *Advantaged*, with high-power level and positively constructed. They receive oversubscribed, overfunded, underregulated, materialized wins, which are also framed as societal wins. Their losses are limited.

2. *Contenders*, with high-power level and negatively constructed. Their wins are primarily hidden, e.g., in legal loopholes. Their losses are highly visible, but difficult to implement and are hence rhetoric (non-materialized).

3. *Dependents*, with low-power level and positively constructed. They receive highly visible, but mostly rhetoric wins. Their wins are fewer compared to the advantaged. Their losses are hidden.

4. *Deviants*, with low-power level and negatively constructed. Their wins are limited or non-existent, while their losses are disproportionate and abundant.

However, the SCPD has limited insights into how technology impacts target populations. In order to do so, we are mobilizing insights from Science and Technology Studies (STS). The STS field suggests that we cannot understand our technologically penetrated society without studying the role of science and technology (Felt et al., 2016). Specifically, STS scholars focus on who benefits from science and technology (Oudshoorn & Pinch, 2008), and question the power and politics of these factors by exploring how the groups are included in decision-making processes that affect them, their access to knowledge and which groups are defined as non-users and thus excluded from access to technologies (Wyatt, Thomas & Tiziana, 2002). STS scholars endeavour to 'open the black box' to insights, by carefully analysing the workings of key sites of science and technology, e.g., Research and Development departments (Hyssalo, Jensen & Oudshoorn, 2016). For these reasons, we aim in this chapter to track how wins and losses are enabled by technology, allowing us to better understand unequal treatment of target groups in smart mobility.

## METHODS

In order to answer our research question, we conducted qualitative research through the study of two behavioural change programmes. We gathered our data via semi-structured interviews and literature review. First, we conducted 12 interviews with decision-makers responsible for implementing the BB programme. Responses of interviewees are shown in brackets in the following sections, e.g., '[1]' referring to interviewee 1.

Semi-structured interviews are a valuable method for obtaining knowledge about the interviewees' attitudes and values towards a specific issue (Byrne, 2004), which is necessary for analysing differential treatment of target groups. Second, we analysed primary policy documents and project websites, which informed us about programme aims as well as about the solutions developed to reach those aims. We analysed the data through applying the SCPD framework, specifically its four target group categories. We mapped our results on these four categories and we analysed the role of science and technology in distribution of wins and losses.

## THE CASE STUDIES

Our case studies, namely the projects 'Burn Fat Not Fuel' (BFNF) in Maastricht and 'B-riders' in Brabant, are part of the national programme Beter Benutten (BB). BB was established by the Dutch Ministry of Infrastructure and Environment (I&M) with two policy goals: to reduce traffic congestion by 20% and door-to-door travel time by 10% during rush hours. To achieve these goals, BB aims to create long-term behavioural change in relation to the daily commute, for example by getting commuters to shift from use of car to bikes or public transport.

One particular intervention to this end, fostered through this programme is the 'connected bike': a(n) (e-)bike with tracking technology that records and provides feedback on the user's activity. One of the connected bike solutions is e-bikes which have a battery that can be used to travel long distances. In contrast with regular bikes, e-bikes are seen by policymakers as similar to the car: It offers an excellent alternative for people using cars for everyday commuting because, as an interviewee suggested: 'the e-bikes are cars without a roof', offering the benefits of motorization to commuters with limited environmental impact and relative comfort.

### Maastricht

Maastricht is one of the oldest Dutch cities, with an internationally recognized university. It is also home to several large companies, employing thousands of residents each, including from nearby cities. According to the interviewees, Maastricht faces daily traffic congestion due to the A2 cross-European highway, crossing the city centre, dividing it in two parts. Moreover, the only traffic light in the A2 highway is located at the entrance of Maastricht, which creates congestion. Following interviewee's utterances, it 'creates a sense of urgency' [4] for policy interventions that improve mobility.

In order to address these challenges, Maastricht established the 'Maastricht Bereikbaar' (MB) organization to run the BB programme in this region. It is characterized by a close collaboration with employers. To be part of MB, employers are required to pay an annual membership fee. This membership approach is considered highly successful for engaging private parties and encouraging them to participate in BB. Based on our interviews, this is the case because it offers policymakers a gateway to reaching their target groups. However, an interviewee suggested that this membership model has the potential to exclude small and medium enterprises, which, due to their size and limited resources, typically lack the dedicated mobility budget necessary to pay these fees. The connected bike programme in the region (BFNF) is mostly aimed at employees with an MB membership. According to the BFNF website, employers in the connected bike programme benefit from: 'healthier employees with lower absenteeism, better accessibility of the company location, lower CO2 emissions, a better company image and savings on parking costs'. From a technological view, participating in BFNF requires a set of two receivers placed on the employee's bike. These receivers calculate participants' daily commuting by bike. At the time this research was carried out, an estimated 1,821 employees have had participated in this programme (Burn Fat Not Fuel, 2016).

To encourage participation, potential participants are offered free-trials of e-bikes and discounts schemes, assistance with purchasing an e-bike (e.g., for the selection of an e-bike model), a motivational coach, as well as bespoke feedback relating to user activity, for example, on users' savings in terms of $CO_2$, fuel costs and health gains from using the bike. Whereas in the previous incarnation of the scheme, the financial incentive was the primary motivation and the feedback secondary; in the scheme's current form the feedback is sold as the primary win. Furthermore, whereas BFNF was exclusive to the car users, the new programme broadens its scope to any employee working in a BB company, based on the assumption that it will build social pressure on car users to switch to bikes. However, due to their specific institutional arrangements of collaborating with employers, in Maastricht only employees working for MB-partnered companies can participate.

## Brabant

The Brabant region encompasses five cities: Breda, Den Bosch, Helmond, Eindhoven and Tilburg. The region is a major contributor of the Dutch national GDP. Its biggest city, Eindhoven, is a hub for technological innovation resulting from an interplay amongst knowledge institutions (e.g., Eindhoven University of Technology) and major technological companies (e.g., NXP and ASML). It is also characterized by having a strong collaboration between actors, especially in the IT sector, and its application to other domains, such as smart mobility (Brainport, 2016).

The multi-city approach in Brabant raises different questions about dealing with the BB goals. In contrast with Maastricht, Brabant focuses not just on inter-city mobility, but also between cities. The structure of BB Brabant differs from BB Maastricht. It is based on a networked organization, with less engagement of employees. Policymakers in this region acknowledged that they have faced difficulties in making employees participate in the programme. The connected bike project in Brabant, 'B-riders', is similar to BFNF, but it does not require any additional hardware to operate. It uses smartphones to calculate distances, provide feedback and offer incentives to participants. B-riders is limited to a maximum number of project participants (5,000).

The B-riders' implementation consists of three phases; of which they are currently in the second one [3 & 6]. In the first phase, car users were offered monetary rewards calculated by a mobility app. This app traced user's activity, rewarding every kilometre using their bike instead of a car (12¢ (<10 km) or 15¢ (>10 km) per km during rush hours, and 8¢ per km outside rush hours). This has been changed in the second and third project phases, where the reward is 10¢ per km only in rush hours and reward points for outside the rush hours. It is only in the later phases that people not commuting by car, but using bikes before the programme was established, started obtaining access to the motivational coach, collecting online points to redeem rewards (e.g., a museum ticket), or entering a lottery (to win a weekend getaway). As one interviewee phrased it: 'to get the money reward you need to be using a car and really change your behaviour' [3].

## FINDINGS

We identified four target groups in both BB programmes: (1) 'modality switchers' – car-using employees working for companies with BB membership, (2) non-car user employees working for companies with BB membership, (3) students and (4) 'traffic participants' – car users not participating in the programme. Although both projects focus primarily on the first two target groups, the interviewees suggested that these projects also target students and traffic participants. The latter group is targeted indirectly: any solution offered to the other three groups will impact car users regardless of their relationship with BB. See Figure 1 for the relationship between the groups and the SCPD framework. The target groups are mostly

**Figure 1.** Social Construction of the Target Groups.

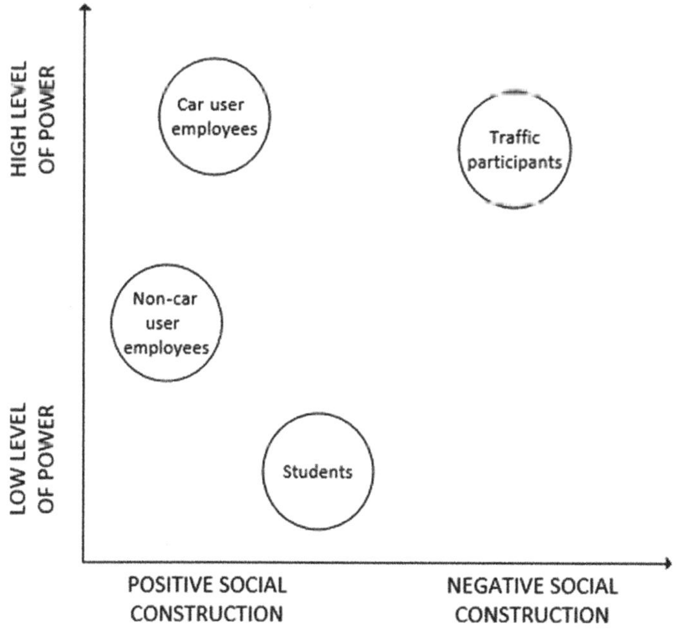

constructed in positive terms. However, their wins and losses differ greatly. Table 1 provides an overview of wins and losses. The differential treatment will be described in the following four subsections.

While answering how technology participates in the distribution of wins and losses, we identified that Ingram and Schneider's (2013) categories of materialized and rhetoric wins are limiting for capturing the wins of the connected bike technology. Instead, we found that certain wins become materialized by ICT, which we call 'digitally mediated wins'. They are more similar to the rhetorical wins, but they become materialized through appearing on a screen of the ICT-personal device, once they are digitally mediated through the use of the connected bike. These wins are non-materialized and putatively non-quantified before the use of the connected bike, because while users could calculate overall their personal fuel costs, $CO_2$ or health savings, we do not expect most users to derive detailed estimations of the health and fiscal benefits they receive from switching from using a car to using a (e)bike. Hence, these promises are more similar to the rhetorical wins as they are only about to become materialized through appearing on a screen of the ICT-personal device, once they are digitally mediated through the use of the connected bike. The smart sensors installed in the connected bike quantify the wins, based on an individual performance of the user. For instance, the health benefit calculation is displayed on a mobile screen. Whereas the Ingram and Schneider's define material wins in terms of fiscal wins, the digitally mediated wins are of different materiality.

## Modality Switchers: Car User Employees Working for BB Members

The daily car users commuting to a BB participating company, who are expected to shift towards the connected bike, have a high level of power, particularly in relation to the other three groups, and are portrayed as the major cause of congestion problems. However, policymakers do not socially construct this group in negative terms. Although we observed that this group is considered as untrustworthy (e.g., they are required to submit a proof of using the car signed by their employer to participate in the programme), they are perceived as a potentially obeying and socially responsible group if the right policy incentives are put in place. Consequently, we

**Table 1.** Wins and Losses per Target Group.

| Target Group | Ideal Type of Target Group | Wins | Losses |
|---|---|---|---|
| Modality switchers | Advantaged | Materialized: e-bike purchasing discounts, lottery, online credit, financial rewards for using a bike; purchasing advice, e-bike trials, online motivational coach, increased health, fuels savings, less stress; common wins | Common losses, social pressures to change their behaviour, no materialized (financial or legal) losses |
| Non-car user employees working for BB members | Between advantaged and dependents | Currently in Maastricht: materialized wins; online motivational coach, online credit, being fit, healthier lifestyles | Currently in Brabant: exclusion from the connected bike schemes; no financial incentives for purchasing e-bikes<br>Currently Brabant and Maastricht: common loses |
| Students | Dependents | Rhetoric wins, i.e., MOOCs, change in lecture times, currently still undefined 'e-bike solutions' | Exclusion from the connected and e-bike schemes; limited access to public transport in rush hours: no financial incentives for purchasing e-bikes |
| Traffic participants: car users (indirect target group) | Contested social construction | Common wins | Common losses, worse lifestyles and health |
| *Common wins and losses* | | Less congested roads, better accessibility of cities | Increased parking fees, narrower roads, more stress, more $CO_2$ pollution, longer travel time |

claim that this group receives a similar treatment to what the SCPD framework suggests as an *advantaged* group: high materialized and rhetoric wins, and no losses.

Policymakers prioritize this group by offering them financial incentives to leave their car at home and to do the daily commute by bike. Next, they received financial incentives and personalized purchasing advice, especially in Maastricht. These incentives were complemented with other benefits, e.g., an online motivational coach. In addition, these wins were easily accessible for the target group. Aligned with the SCPD framework, policymakers consider that these type of wins are societal wins as it is in the best interest of both regions to induce long-term behavioural change in this group. Additionally, policymakers perceive that this group's participation in the programme can improve their quality of life and health, framed as the two main incentives for this group to adopt the connected bike. Yet, this group has limited losses. The limited symbolic punishments are social pressures to change their behaviour, without materialized losses, e.g., no fiscal punishment.

### Non-Car User Employees Working for BB Members

Employees who are non-car users, working for BB members are positively perceived by policymakers because they already commute through more sustainable modes than the car. Policymakers, therefore, aim to reinforce their behaviour and discourage them to shift to cars for daily commuting. However, we see limited financial rewards for this group, and their benefits are less materialized than for the modality switchers. We argue that this group, therefore, fits between the categories of *advantaged* and *dependents* within the SCPD framework, implying a mix of rhetoric and materialized wins, but less materialized than the advantaged group of the first identified group, and fewer losses than the other three groups.

This group has been treated differently in the two regions, and throughout the different phases of the BB project. In the early stage of the project, Maastricht excluded this group from any wins while in Brabant it was offered online motivational coaching, credit and participation in a lottery. In the later stages, policymakers in Brabant took out these benefits. When excluded from the above-mentioned wins, this group does not have any preferential treatment; thus, we consider them as the overall loser.

Several policymakers highlighted that keeping this group using the connected bike is, rather than a goal itself, a policy instrument that can be used to influence behavioural change in the other groups.

### Students

Although interviewees acknowledged that students are a target group of BB, they benefit from few operationalized measures. Often, interviewees suggested materialized wins which are expected to be implemented in the near future, but up until now had not realized. We observed a tendency of prioritizing the other three target groups at the expense of students. For example, policymakers aim to change the behaviour of students commuting to university by taking them out of rush hours, allowing employees to benefit from less-congested transport infrastructure. We identify a social construction in which policymakers see students as a *dependent* group in the SCPD: a positive social construction, but with limited power to access and participate in the connected bike initiatives.

According to policymakers, students are not eligible for participating in the e-bike solutions as they cannot afford them. This group is considered by policymakers as predominantly bike users; however, limited attention is given to measures for reinforcing this behaviour. Additionally, policymakers consistently refer to materialized wins for students in the near future, such as discounted e-bikes. However, during the time of our research, these were never materialized. Students do not have access to any financial incentive to switch their behaviour. They do not have access to other solutions such as to e-bike buying schemes. An additional argument of policymakers to neglect this group is that it already has concrete mobility wins, such as free access to public transport. Furthermore, students are considered to complicate the behavioural change of other groups as they use mobility infrastructure (public transport) during rush hours. The major tool used to induce behavioural change is reducing students' mobility during peak hours. Thus, policymakers offer students solutions such as Massive Online Open Courses. However, we perceive this type of solutions as a burden for students, as their main intention is to limit their mobility. We identify students as the overall losers in our case-study.

### Traffic Participants: Car Users Not Participating in the Programme

The traffic participants' group consists of daily car commuters who do not participate in the BB programme. Unlike the other groups, they have a contested social construction, indicating that policymakers have a neutral view on this group. However, they remain a powerful group as policymakers acknowledge that getting traffic participants to stop using their car without any incentive is difficult, so their 'bad behaviour' is not punished by any means. Policymakers suggested that this group has high rhetoric losses. For example, policymakers state that this group will be affected as their continued car commuting results in worse lifestyles and health. However, there are no materialized losses. This group may access the most symbolic wins provided by the BB programme, such as rewards points or an online motivational coach. Because this group benefits indirectly from the connected bike initiatives, we consider them also a target group.

## CONCLUSION

Returning to the question of 'what is the role of technology in the differential treatment of target groups in the BB connected bike projects in the regions of Maastricht and Brabant?' we identified four groups. Each group receives differential treatment in policy, which is primarily targeted via incentives to encourage connected bike adoption. We found that the 'modality switchers' group, which contributes most to the problem of congestion and travel time in the Netherlands, is also receiving the highest number of material wins in the connected bikes projects of Maastricht and Brabant. In contrast, other groups, such as students and people cycling before the programme started, are neglected or attributed a reduced number of wins. Additionally, traffic participants benefit primarily from indirect wins. This policymaking approach may hinder the transition towards smart mobility technologies, and limit the emergence of alternative transportation modes to lower congestion and sustainability.

We propose that Dutch attempts towards stimulating the adoption of the connected bike should embrace a broader focus, including other groups. Moreover, we found it striking that the connected bike projects lack clear losses to groups which are contributing to the problem the most. A more balanced approach in the policy design should be to adopt measures for punishing negative behaviour, hence the car users. Furthermore, it should aim for more symmetrical approaches and supporting a more inclusive distribution of the wins.

Furthermore, policymakers are increasingly shifting responsibilities for the social construction and preferential treatment of target groups to private parties, such as employers. This raises questions about equity and justice in the policy design of smart mobility technologies. Market parties are incentivized to participate in areas in which profit is reachable. However, some groups (e.g., students) would not typically fall into these groups. This uneven treatment of target groups in smart mobility may justify public intervention, in order to maximize the benefits of smart mobility technologies to underserved populations. ICTs allow individualized tailoring of treatments, which implies a high need for policy intervention in setting ground rules, to ensure ICT solutions are consistent with supporting outcomes which are both fair and sustainable. Incentivization and reward alone will not resolve the market and operational failings that characterize the system today.

## ACKNOWLEDGEMENTS

We thank the Dutch Ministry for Infrastructure and the Environment and Rijkswaterstaat for funding this research project. We also thank the participating interviewees, as well as Sven Nyholm and Geert Verbong, for their fruitful comments.

## REFERENCES

Brainport. (2016). Co-creating the future. Retrieved from http://www.brainport.nl/en/about-brainport/co-creating-the-future. Accessed on 18 May 2016.

Burn Fat Not Fuel. (2016). Burn fat not fuel. Retrieved from http://www.burnfatnotfuel.nl/. Accessed on 18 May 2016.

Byrne, B. (2004). Qualitative interviewing. In C. Seale (Ed.), *Researching society and culture* (pp. 179–192). London: Sage Publishing.

Felt, U., Fouché, R., Miller, C. A. & Smith-Doerr, L. (2016). *The handbook of science and technology studies*. Cambridge, MA: MIT Press.

Geels, F. W. (2012). A socio-technical analysis of low-carbon transitions: Introducing the multi-level perspective into transport studies. *Journal of Transport Geography*, *24*. 471–482.

Geels, F. W., Kemp, R., Dudley, G. & Lyons, G. (Eds.). (2012). *Automobility in transition? Socio-technical analysis of sustainable transport*. Abingdon: Routledge.

Hyysalo, S., Jensen, T. E. & Oudshoorn, N. (2016). *The new production of users*. Abingdon: Routledge.

Ingram, H. & Schneider, A. (2013). Social constructions in the study of public policy. In J. A. Holstein & J. F. Gubrium (Eds.), *Handbook of constructionist research* (pp. 188–211). New York, NY: Guilford Publications.

Ingram, H., Schneider, A. L. & Deleon, P. (2007). Social construction and policy design. In P. Sabatier (Ed.), *Theories of policy process* (1st Ed., pp. 93–127). Denver, CO: Westview Press.

MacKenzie, D. & Wajcman, J. (1999). *The social shaping of technology*. Buckingham: Open University Press.

Oudshoorn, N. & Pinch, T. (2008). User-technology relationships: Some recent developments. In E. J. Hackett, O. Amsterdamska, M. E. Lynch & J. Wajcman (Eds.), *The handbook of science and technology studies* (3rd Ed.).

Pierce, J. J., Siddiki, S., Jones, M. D., Schumacher, K., Pattison, A. & Peterson, H. (2014). Social construction and policy design: A review of past applications. *Policy Studies Journal*, *42*(1), 1–29. doi:10.1111/psj.12040

Schneider, A. L. & Ingram, H. M. (1993). Social construction of target populations: Implications for politics and policy. *The American Political Science Review*, *87*(2), 334–347. doi:10.2307/2939044

Wyatt, S., Thomas, G. & Terranova, T. (2002). They came, they surfed, they went back to the beach: Conceptualizing use and non-use of the internet. In S. Woolgar (Ed.), *Virtual society?* Oxford: Oxford University Press.

# SECTION THREE

# STATE CAPACITY

# CHAPTER 7

# GOVERNMENTAL CAPACITY AND THE SMART MOBILITY TRANSITION

Diane E. Davis

ABSTRACT

*Using materials drawn from San Francisco and Stockholm, this chapter assesses the extent to which recent efforts to upgrade transport services through smart mobility technologies have advanced short- or long-term urban policy aims in the arena of transport governance for sustainability. We argue that positive governance impacts depend largely on degrees of coordination and oversight. Our findings suggest that these aims are not going to be easily met by a network of competing private firms or individuals using smart technology to achieve their own singular trip priorities. Stated in the lingo of social science, the smart mobility transition will produce a 'collective action problem' if it remains in the hands of individual firms without some larger territorial and service coordination by governing authorities. To counter this possibility, we argue that transparent implementation processes involving multiple stakeholders will offer the best opportunity for ensuring that smart technology innovations will become a means for expanding governance capacity.*

**Keywords:** Institutional capacities; smart technology; ridesourcing; congestion pricing; implementation processes; stakeholder coordination

## DISRUPTION IN THE TRANSPORT SECTOR

Innovations in the IT sector have begun to transform transportation supply and demand in cities around the world (Batty et al., 2012; Meijar & Bolívar, 2016). Yet it is important to recognize that advocates of the smart mobility transition do not always concern themselves with governance or coordinating capacity, let alone a larger commitment to advancing long-term urban sustainability aims. In the current era, promoters of new or improved urban transportation services, IT-related or otherwise, are often focused on the difficult task of challenging the primacy of the automobile and expanding non-motorized mobility alternatives and reducing vehicle congestion in whatever way possible. Any policy that does so is likely to be embraced with considerable enthusiasm, and doing so will inevitably be a first and important step in establishing the pathways for change, and in this regard should be considered a transport policy success. Still, the question that concerns us here is not merely policy success but also governance capacity. For this reason, we ask whether some smart mobility innovations might lay more fertile groundwork for strengthening coordination mechanisms across multiple sectors, levels of government and territorial scales as well as altered planning institutions, public–private sector relations, civil society commitments or governance mandates.

To address these concerns, we consider IT-enabled ridesourcing services (Millard-Ball, 2016). Ridesourcing has been controversial because it frequently challenges long-standing industry patterns, particularly in contexts where taxi drivers are well-organized and protected by labour protections or livery service regulations (Dudley, Bannister & Schwanen, 2017). Some observers have gone so far as to characterize ridesourcing as a rapacious revenue-generation model masquerading as a new mobility paradigm (Söderström, Paasche & Klauser, 2014). Yet the value of IT-led innovations in transport must be judged by criteria other than their profit-making potential. The issue is as much the operation and management of such innovations as it is the nature of servicing (Guerra, 2016). Indeed, evidence suggests that although the application of smart technology to livery services may afford greater accommodation of individual service demands, such innovations also hold the potential to undercut demand for public transport, particularly mass transit services, thus producing financial barriers to adequate service delivery in the public sector more generally.[1]

All this is not to say that there have been no gains. A wide range of smart mobility technologies are being effectively deployed by public transport agencies to help better service, whether in the form of real-time communication about schedules and vehicle availability, or even to monitor and manage congestion and demand on both rail and roadway. Accordingly, the larger implications of the smart mobility transition are still not entirely clear, particularly because IT-sourced transit services can either complement or compete with public transport operations, depending on how they are deployed, by whom and in what context. Likewise, the growing fascination with autonomous vehicles could produce a future where city streets post a wide array of vehicles smart enough to avoid clashes but not necessarily wise enough to stop driving when the aggregate number of trips or vehicles reaches a certain urban congestion threshold. To avoid such results would require governing or regulating institutions charged with operational oversight of an entire network, a task that cannot be left to the vehicle itself. Smart technology may allow vehicles to communicate with each other, but in order to advance aggregate transport and mobility gains someone needs to be monitoring and coordinating the system as a whole (see Chapter 5 for an expanded discussion on autonomous vehicles).

In this chapter, we ask whether the smart mobility transition has affected the capacities of governing authorities to advance environmental sustainability, compact city form and other related priorities historically linked to effective and equitable transport. We consider whether recent efforts to upgrade transport services through smart mobility technologies has been enabled by different governance strategies and tactics – and if so, whether these strategies and tactics themselves have affected the achievement of short- or long-term policy aims in the arena of transport governance for sustainability. Our efforts to link the smart mobility transition to expanded governance capacities parallels one of the most recent discursive evolutions in the transport policy world, which is the gradual shift in focus from solving problems of transit to enhancing accessibility (IDRP). The former is defined in terms of trip times, trip length and the structure of routes: that is, how persons or vehicles move from point A to point B. In contrast, the latter takes into account the entire network of transit options, asking how well such a system serves citizens, and considering which ones in particular are advantaged or disadvantaged by existing transit options. With this intellectual shift to examine more than the technical aspects of mobility, we join other transport scholars in seeking to jettison the narrowly cast preoccupation with transit services and their technical operation, and begin exploring questions about overall mobilities as well as how to maximize access for all – albeit with a focus on the smart technology transition and its relationship to oversight, coordination and the expansion of transport governance capacities more generally.

## TRANSFORMING GOVERNANCE CAPACITY: POLICY VERSUS PROCESS?

To assess the relationship between the smart mobility transition and governance capacity, this chapter draws on material from two cities, which have adopted new technologies for monitoring, management and/or provision of transport. These are San Francisco's early adoption of an accommodative regulatory framework to support the proliferation of ridesourcing services (i.e., Uber and Lyft) and Stockholm's adoption of a surveillance technology-monitored congestion charging system, whose revenues financed local and regional transport infrastructure investments. We seek to ascertain whether the roll out of smart mobility components within the larger set of transportation changes pursued by local authorities in these two locations has positively or negatively impacted cross-agency coordination, multi-scalar agreements, public–private relations and other governance factors fundamental to enhancing each city's capacities to achieve sustainable transport aims.

We also want to know *how* and *why* any such changes may have occurred: Is it something about the smart mobility policies themselves and how they transform the transport network? Or, does it owe to the implementation process, that is, to the manner in which IT innovations in transport are introduced – ranging from a focus on who is involved to the strategies and tactics their advocates employ? In particular, we are interested in the number, sectoral and territorial range of stakeholders, their singular or combined political clout, and the hierarchical nature of the relationships among them, including whether alliances are made horizontally or vertically across sectors or domains and whether innovation is driven by state actors, civil society, the private sector or some combination. This gives us a general sense of process as well as leadership strategies and tactics. In documenting the main actors and institutions advocating these smart mobility policies, we also seek to identify key aspects of the implementation process, including degrees of contestation, coordination and oversight.

After laying out these general dynamics, we compare the two cases, asking whether there exists any relationship between a smart mobility policy's mobility impacts, implementation processes and enhanced governance capacities. We will define enhanced governance capacity with a view to four dimensions: (1) introduction, reinforcement or expansion of sustainability objectives within existent governance mandates and institutions, (2) introduction, reinforcement or expansion of sustainability objectives within civil society, (3) the establishment of new networks of coordination to connect cities with other actors and institutions across multiple territorial and sector scales of governance and (4) greater collaboration between public and private-sector actors around sustainability goals. We then offer some concluding generalizations about stark differences between the two cases.

## RIDESOURCING IN SAN FRANCISCO

Probably the most-debated disruptive innovation in transport servicing in the past several years has been ridesourcing, defined as a means of servicing customers through phone applications that leverage real-time mapping of locational demand. Industry leaders in this field actually launched their efforts in the city of San Francisco, where the emergence of privately owned but publicly regulated companies such as Uber, Lyft and Sidecar dramatically changed not just the local market but also the global urban transportation landscape, most notably by altering the expectation (and reality) that routine urban transit services can be readily tailored to disaggregated individual preferences. As of 2015, after only four years of legal operation, ridesourcing companies in San Francisco were providing about 50,000 trips a day, more than twice the number of trips taken by taxi.[2] These dramatic changes are largely the creation of tech-savvy private-sector actors, who developed the necessary applications and business models that kept costs low by relying on drivers using their personal cars and personal car insurance.

However, to enable these changes, tech companies also needed an accommodating regulatory framework, not just in the face of opposition from the city's heavily monitored, politically active taxi operators but also under the continued threat of regulatory crackdowns by local and state enforcement agents. Stated differently, the successful adoption of ridesourcing was all about changing governance. In finding room for manoeuvre, ridesourcing firms gained support from San Francisco Mayor, Edwin M. Lee, who took office in 2011 as the 'pragmatic' candidate intent on bridging the progressive and moderate factions of the Board of Supervisors. In a city still feeling the impacts of the Great Recession, Lee publicly supported ridesourcing as an embodiment of the 'sharing economy'. In order to circumnavigate local enforcement requisites, Mayor Lee shifted the jurisdictional locus of debate concerning regulatory activity to the state level – specifically the California Public Utilities Commission (CPUC), which was generally viewed as a more business-friendly venue and where taxi interests typically have less influence. In turn, CPUC Chairman Michael Peevey convinced the five-member commission to begin a rule-making process focused on ridesourcing. In fall 2013, CPUC approved regulations which would be applied to a new category of service, 'Transportation Network Companies' (TNCs), which imposed obligations related to insurance, background checks for drivers and vehicle inspections but did not require substantial changes in existing business models. These new rules and the manoeuvring that produced them allowed ridesourcing to continue and grow in California. Their impact has been so great that both cities and technology firms worldwide have since then sought to replicate similar regulatory modifications, albeit with varying degrees of success.

The introduction of ridesourcing owed to a combination of leadership strategies and political tactics; although in this case, private-sector firms initiated the process by engaging public sector actors in a shared effort to challenge and eventually alter a series of long-standing regulatory frameworks governing personal transportation services. This outcome was facilitated not just by Mayor Lee's rescaling of key decision-making authority about transit regulations from the local to the state level, where opposition forces had less traction and where technology firms held more sway. Other strategic priorities, such as retaining and wooing technology companies involved in the sharing economy, explain his openness to the initial lobbying from Uber and Lyft. For their part, ridesourcing firms used email and social media to mobilize their existing user and driver base as well as potential supporters, urging them to contact elected officials, attend public meetings and sign online petitions in support of ridesourcing services.

Tactical measures included aggressive and forceful rejection of regulatory restrictions, both behind the scenes and in the public realm. The leaders of ridesourcing firms initially argued that they weren't subject to city or state regulations and could begin operating without regulatory approval since they were only facilitating connections between riders and drivers (see Chapter 4 for the importance of working around definitions). When met with regulators' warnings and orders to cease operations, they simply ignored them while trying to build support and see if regulators would actually act against them. They also brought public sector allies to their side by arguing that private provision of surface transportation services could be a means for alleviating pressures on public transit coffers. In short, ridesourcing in San Francisco was a predominantly, unsubsidized initiative led by technology-savvy private-sector actors, who developed the necessary applications and business models that kept transport costs low by relying on drivers using their personal cars and personal car insurance.

All this is not to say that the smart mobility policy advances in San Francisco did not address some basic transport network needs. Service failures of the local taxi industry were notorious, and help explain the lack of organized public opposition to ride-sharing or subsequent efforts to change state regulations. All this further meant there was very little open public conversation or contention about the regulatory changes. The end result was a relatively rapid and surprisingly conflict-free process in which smart mobility innovators were able to introduce a cutting-edge mobility service capable of tailoring urban transit services to a large urban market. Yet they did so by using a tactically incremental approach, first arguing that they weren't subject to city or state regulations and could begin operating without regulatory approval, then ignored regulators' warnings and orders to cease operations while trying to build support and see if regulators would actually act against them, and finally tried to convince elected and senior appointed officials to support the development of new, more favourable regulations for ridesourcing in the face of crackdowns. In the face of these aggressive actions, formal decision-making by state regulators was somewhat hasty, top-down and opaque.

## CONGESTION CHARGING IN STOCKHOLM

In August 2007, Stockholm introduced a congestion charge for cars crossing the city's inner boundary, aimed at reducing traffic flows into central city areas. A central technological component of the policy involved the use of surveillance cameras to monitor and bill vehicles in movement. The decision followed a seven-month trial from January to July 2006 and a public referendum that followed just seven weeks later, coincident with both national and city elections. Today, congestion charging is not only a centrepiece of Stockholm traffic management; it is also a major source of revenue for transit and road investments throughout the greater Stockholm region – possible primarily because the charge is considered a national tax and not a local charge.[3]

Strictly speaking, the deployment of surveillance cameras in Stockholm's congestion charging scheme may seem to be a relatively rudimentary application of smart mobility technologies for the purposes of bettering transit, at least when compared to ridesourcing. Yet the cameras' capacities to monitor traffic volume without reducing vehicle speed were an important part of the program's success, as were the algorithms formulated to measure and monitor traffic flows. Likewise, installed cameras allowed both a transparent accounting for and a real-time visual display of new traffic flows, both of which were critical in generating voter confidence for the measure. From the moment of the trial, local newspapers published pictures that showed the smooth operation of the system, and such evidence further strengthened public support for this policy. In addition, metrics on time saved and the amount of public revenue generation both derived from and revealed the value of monitoring the volume of cars entering the city. In many ways, one could say that the success of this highly controversial policy owed to the advanced technologies employed in its operationalization.

But just as important were the strategies and tactics used to promote and implement the policy. The mayor behind this program, Annika Billström of the Social Democratic party, actually campaigned for office on a platform of opposition to congestion pricing, primarily because of her concern about losing conservative automobile-oriented voters in the suburbs. When she assumed office, however, she was pressured by the local Green and Stockholm Parties as well as her own Social Democratic party colleagues at the national level to nonetheless move forward on congestion charging against her own campaign promises. One way that Mayor Billström 'saved face' was to introduce a pilot rather than a full-blown policy change. She also actively managed the technological logistics of policy implementation by bringing key private-sector actors in IT on board in the preparation for the trial, including high-profile technology companies such as Eriksson and IBM.

More generally, Mayor Billström pursued a range of strategies and tactics to help turn a potential political humiliation into an opportunity to put congestion charging back into the realm of possibility. With a skilful management of the trial, which itself directly led to the successful referendum, she showed the public that there was a way to restructure overall transit and mobility patterns through a policy that was anything but disruptive, and indeed publicly palatable. Perhaps most important, Mayor Billström introduced congestion charging to the public not as a done-deal mandated by others in the Social Democrat–Green coalition without concern for Stockholm, but rather, as an experiment on which local citizens could vote and have the final say on their city's transport future. She reconceptualized the policy in ways that would minimize contention and partisan criticism. That is, she kept open the possibility that a permanent congestion charging policy might ultimately be rejected by a majority of voters, thus helping solidify public support by introducing elements of democratic accountability to the public.

Last but certainly not least, the mayor openly recruited and empowered a technical team to package transport knowledge in a politically savvy manner. In advance of the trial, for example, transportation planners and administrators made significant expansions in bus service and new subway cars to indicate to voters that they were 'getting something' from the revenues generated by the charge. During the trial, the city also ran a public outreach and education campaign that communicated in clear and accessible terms the trial's purpose, workings and benefits. And after all these careful and deliberate moves – intended to pre-empt opposition and/or technological failure – the public accepted the change. Her successes were then immediately seized upon by her original opponents, who in turn highlighted congestion charging's potential to generate revenues for a wide range of transit and urban development investments. Such a reframing facilitated subsequent inter-party and regional consensus necessary to gain parliamentary approval for the congestion tax, thus expanding the range of political stakeholders in ways that ultimately would widen institutional commitments and expand overall governance capacity.

To summarize, Mayor Billström relied on a series of strategic and tactical actions to produce public sector, private sector and citizen embrace of congestion charging in Stockholm. In addition to the points noted earlier, they included: (1) introducing the trial at *full scale* – with as few exemptions as politically feasible – so as to demonstrate significant impact as rapidly as possible, (2) ensuring the technical proficiency of the entire operating system by engaging politicians, planning professionals and private-sector partners from early on and proactively troubleshooting together as needed and (3) overcoming coordination and implementation obstacles by convening working groups across departments and with direct reporting channels. In and of themselves, any one of these strategies or tactics would neither be surprising nor precedent-setting if it weren't for the uphill battle the mayor faced from the get-go. Although Mayor Billström's successes after so many years of failure could have been advanced by the accelerating environmental consciousness and activism in Sweden, the adoption of congestion charging in 2007 must be viewed as the result of more than the changed temporal context and growing concern about the climactic dangers of growing automobile emissions. It also emerged out of the ashes of conflict unfolding over several decades that was finally overcome through negotiation and compromise involving different political parties, a wide range of technical and bureaucratic actors creating horizontal relationships among themselves at the level of the city, the direct participation of multiple governing actors and institutions at the local, regional and national scales and the strategic leadership of the mayor herself.

## POLICY SUCCESS, IMPLEMENTATION PROCESSES AND GOVERNANCE OUTCOMES: COMPARING SAN FRANCISCO AND STOCKHOLM

So how should we compare the cases of San Francisco and Stockholm? In both cities, private-sector actors involved in IT worked with their respective mayors to help advance the adoption of smart mobility applications in transport servicing, although there were key differences in how this unfolded. In San Francisco, the private sector led much of the collaboration, whereas in Stockholm technology firms were brought on board after the decision to move forward with congestion charging was finalized. More significantly, San Francisco's private-sector leadership was tied to a single mode of transport and it was focused on individualized public transportation solutions (i.e., ride-sharing). In Stockholm, private-sector involvement was closely connected to public sector leadership, which itself was not only directed towards multiple transit priorities (i.e., it was multimodal), but also was more 'social', in that the public and private-sector collaboration revolved around the discussion of mass transit and public roadway investments – territorialized at the regional scale, and the shaping of urban form and expansion generally.

All this suggests that if we were to think of these two cities' smart mobility innovations on a more transformative spectrum, identifying the potential of each to achieve long-term mobility goals through enhanced governance capacities as opposed to merely advancing short-term mobility goals through one-off policy successes, Stockholm and San Francisco might be considered on two opposite ends of the scale. In Stockholm, congestion charging both reduced the flow of carbon emitting automobiles while also generating substantial revenues that were fuelled back to government administrators for investment in other sustainable transport goals. In San Francisco's ridesourcing, the fiscal gains accruing from this smart mobility innovation remained in private-sector hands, and its impacts were meaningful for individual transit consumers but less so for local authorities. Overall, there were few positive spillovers for the operation of the entire transport network, either in budgetary or transit coordination terms.

To be sure, it would be foolhardy not to recognize that ridesourcing is a hugely transformative industry for the provision of mobility services, and that without the success in San Francisco, the transit community might not be seeing the global expansion of ridesourcing services. But if we were to look at the institutional and governance arrangements altered by its introduction, at least in the San Francisco case, we might think differently about where its transformative impacts were most or least felt. Indeed, one would be hard-pressed to identify institutional or governance modifications that accompanied or were set in motion by the adoption and expansion of ridesourcing in San Francisco. The main changes were confined to regulatory modifications directed only towards this particular service and how it might fill gaps produced by an unresponsive taxi industry. No new institutions were developed to manage the changed regulatory environment, nor were other city or state agencies invested in overall transport planning, given additional mandates to extend or integrate newly offered ridesourcing services into their governance mission.

Just as important, the new regulatory changes adopted in San Francisco were not clearly presented as being central to an integrated network of sustainability priorities intended to create synergies between public and private transit, strengthen transport affordability, streamline transport–land use coordination or reduce congestion. Private-sector promoters of ridesourcing were primarily interested in making it possible for their vehicles to circulate without constraint. As such, they did not see their role as buttressing the priorities of the city's transport planning institutions in any obvious way. The mayor, for his part, framed support for ridesourcing with his eye on the city's commitment to high-tech innovation and the ways in which old livery regulations would prevent the rolling out of these new mobility products. This not only meant that he was less embroiled in conversation with city transport planners than with state code regulators, but also that he appears to have been more driven by concerns about the city's reputation and branding as the nation's high-tech capital than he was about addressing San Francisco's sustainable transport challenges through these new measures.

In Stockholm, in contrast, congestion pricing not only produced revenues that helped spur larger changes in public transit systems, but also produced a range of transport and land-use planning commitments from political leaders at different levels of government, in ways that strengthened connections between a variety of public and private-sector actors, rather than vice versa (as in San Francisco). To be sure, it is reasonable to suggest that such differences between San Francisco and Stockholm owed to the policies themselves, with congestion charging by its very nature having more transformative governance potential to expand capacity because of its revenue-generation dimensions. That may be partly true. But ridesourcing also holds the potential to advance sustainable transport goals, depending on how it is coordinated or managed and by whom. Since its much-heralded beginning in San Francisco, built around the 44 regulatory shifts that made such services legal, ride-sharing regulations have continued to evolve even in that city, and some are even being used to address the last mile problem in combination with public transit. Many now argue that the next step will be to promote more pooled ridesourcing services as well as multimodal integration that could include drop off at transit stops as well as improved access for underserviced populations.

That so little of this happened in San Francisco owes to the nature of the implementation process, in which smart mobility innovations were advanced by private transit providers more than public officials or city transportation planners. In San Francisco, moreover, new ridesourcing regulations were approved with the involvement of a relatively small number of actors confined mainly to a small number of regulatory agencies. Debate over the purpose of the policy and its subsequent implementation was depoliticized, closed, narrow and technical – unfolding behind the scenes in meetings of state regulators. This meant that there was very little public discussion about the pros and cons of ridesourcing even generally, let alone in light of city's larger sustainability aims and its regional transit system. In Stockholm, in contrast, from the beginning, the process involved governing actors from at least two scales (local and national), whose support for congestion charging with a subject of open public discussion and even criticism. Moreover, the process in Stockholm was openly contentious, bumpier and incorporated transport and regional planners as well as politicians from different parties, all of whom did not necessarily agree with each other but who were committed to open dialogue. In consequence, the pros and cons of congestion pricing were subject to constant citizen scrutiny, and a wide range of stakeholders had the opportunity to weigh in on their views – not just in the public realm, but also in an open electoral referendum. Ultimately, when the public signalled its support of the policy, there already existed a track record of commitments to linking the program to sustainability aims.

Given these differences, it is perhaps no surprise that Stockholm has continued to advance its trajectory as a sustainable city, built on concerted efforts to manage traffic, to connect housing, transport and land use and to use IT in achieving these aims. Just a few years after the introduction of the congestion charge, Stockholm was named Europe's sustainability capital. San Francisco, for its part, has the IT brand. But it still has not been recognized in any significant way in global transport or sustainability circles as being a transit leader.

## CONCLUDING REMARKS

Advancing universal claims or attempting to draw robust conclusions from the comparison of two cities known for adopting quite different smart mobility innovations in the field of transport is bound to be fraught with problems. In addition, San Francisco and Stockholm are notably different cities in terms of political institutions, governance arrangements and structures of decision-making – having a lot to do with the nature of their democracies. Stockholm is the capital of the welfare state, with a robust political party system and a history of engagement between national and local authorities. San Francisco is an outpost city in a federal system characterized by historical commitments to decentralize democracy in which cities often adopt policies without national intervention, and sometimes even without oversight from their respective state governments. In the United States, moreover, the balance of state and market is quite different than in Sweden, where active government involvement in transport servicing, land use planning and urban policymaking is the rule more than the exception (see Chapter 8 for further discussion on Australia which is more aligned with the US balance). For that reason, the efforts of private-sector IT actors to go it alone in San Francisco and their counterparts' willingness to collaborate with governing authorities in Stockholm are not surprising.

It is not our objective in this chapter to take a normative position on either system, so much as to look more carefully at what makes it possible to expand governance capacities through smart mobility innovations. At first glance it appears that some of this has to do with the policies themselves. Congestion charging can readily provide opportunities for expanding governance capacity by virtue of its revenue-generation potential – although, the selection of certain policies can also reflect differences in political systems.

Even so, we would like to suggest that expansion of governance capacities is a function of implementation processes as much as any particular policy preference or larger political context.

We have seen that convening multiple actors across levels of government, parties and sectors to disagree, debate, negotiate and build consensus about the value of a given smart mobility innovation can help institutionalize organizational commitments to change beyond a single policy success. Complex and contentious implementation processes that involve multiple stakeholders who wrangle over a certain policy outcome will not only enhance the number of institutional players with 'skin in the game' of sustainability, but also can create more transparent dialogue across various constituencies, opening up ways for further engagement around sustainability issues in the long term. Implementation processes that bring transit providers into deliberation with public officials and private citizens will ensure that the adoption of smart mobility measures is more likely to fulfil expansively-defined urban aims rather than narrowly-cast mobility objectives associated with either the production *or* consumption *or* management of transit services.

Farsighted and proactive leaders also make a difference. Within a given network of stakeholders, those capable of articulating a larger urban vision – within which a certain smart mobility innovation might be adopted – play a particularly important role in ensuring transformative change. Sometimes, a particularly conscientious, strategic or forward-looking mayor will play this role. Yet urban planners have historically taken on such commitments: identifying a desirable urban future, and then formulating the implementation measures needed to get there. Thus it may be no surprise that in Stockholm – where implementation processes involved the largest number of players and where smart mobility measures were framed in the context of larger urban objectives leading to expanded governance capacity – both the mayor and urban planners helped formulate and monitor congestion policy rollout. The direct involvement of urban planning professionals in Stockholm may seem to reflect differences in political systems, because planning in welfare state conditions is the norm, whereas in governance contexts where market dynamics rule and state intervention is viewed suspiciously, planners are often powerless or delegitimized in major policy decision-making (see Chapter 8). Yet it must be noted that planners are not entirely absent in advancing smart mobility innovations in all American cities. As such, the variations in expanded governance capacity discussed here are more a product of specific differences between San Francisco and Stockholm, more than the United States and Europe were generally.

Perhaps this last point is the most important. Today we are barraged with nonstop, disruptive innovation in the IT sector, a trend clearly associated with the city of San Francisco and the iconic Silicon Valley that surrounds it. Transport is merely one domain in which disruptive practices associated with firms in the San Francisco area are challenging standard industry practices. And few would doubt that the transport industry in its entirety is being turned upside down through its continually ramped-up efforts to remake itself through IT, and doing so in the form of new collaborations with IT firms in San Francisco and elsewhere. This, however, may be cause for alarm as much as celebration. City leaders and smart mobility experts alike should not lose sight of where cities could be heading under the pressure to enable IT industry changes. The cases discussed here suggest that multiple paths could be followed. Linking smart mobility innovations to future visions of the cities we want to live in can serve as a first step towards articulating sustainable transport goals; whereas the failure to think proactively may stymie opportunities to expand governance capacities for actually achieving such aims.

It is time to think long and hard about what types of cities we are producing with the latest generation of mobility innovations. We must be prepared to frame discussion of transit improvements – smart mobility inspired or otherwise – in light of their impacts on governance capacities to monitor and guide sustainable city form. We should be asking citizens and their governing authorities to think about the kind of city they want to live in, and then adopt or advocate for the mobility systems most likely to lead us there. We should not succumb to technological determinism and let our future cities be the aggregated product of a series of random innovations in the IT industry, appealing as they may at first appear. Rather, we must reverse the order of change: imagining the urban conditions that make us human as well as socially and environmentally sustainable as a species, and then finding and promoting the policies and innovations that help achieve those aims. Otherwise we are merely lemmings running towards the sea, following the promise of a few appealing mobility innovations without seriously considering where they will ultimately lead us to.

## NOTES

1. In addition to reducing demand for public mass transit, there also are indications that a proliferation of ridesourcing services may increase street congestion rather than decreasing it, owing to the fact that ridesourcing serves as an opportunistic mechanism for part-time employment.

2. Uber was founded in 2009 and launched its beta version in 2010. For more on the metrics of ridesourcing successes as well as the politics of its implementation in San Francisco, see Flores-Dewey and Rayle (2018).

3. For a complete account of the politics of congestion charging in Stockholm, see Davis and Radar Olsson (2018).

## ACKNOWLEDGEMENTS

This chapter builds on findings generated by a three-year research project focused on the role of political leadership in transforming urban transport, funded by the Volvo Research and Educational Foundations (VREF). I was the project's Principal Investigator (and a field researcher for the Stockholm case). Primarily our research focus was on leadership strategies and tactics used to transform urban transport in eight cities around the world. In this chapter, my focus is the smart mobility transition and its impact on governance capacities. The analysis presented here is entirely my own, and does not represent the views of the larger TUT-POL research team or any individuals within it. I owe a special debt of gratitude to Onesimo Flores Dewey, who conducted the primary research for San Francisco case discussed here, as well as Lisa Rayle his co-author. Thanks also to the project's Senior Research Associate, Lily Song, who offered considerable wisdom and assistance in the crafting of this chapter.

## REFERENCES

Batty, M., Axhausen, K. W., Giannotti, F., Pozdnoukhov, A., Bazzani, A., Wachowicz, M. & Portugali, Y. (2012). Smart cities of the future. *The European Physical Journal Special Topics, 214*(1), 481–518.

Davis, D. & Radar Olsson, A. (2018). Congestion charging in Stockholm. In Davis & Alshuler (Eds.), *Transforming urban transport*. Oxford University Press.

Dudley, G., Bannister, D. & Schwanen, T. (2017). The rise of Uber and regulating the disruptive innovator. *The Political Quarterly*, *88*(3), 492–499.

Flores-Dewey, O. & Rayle, L. (2018). Ridesourcing in San Francisco. In Davis & Alshuler (Eds.), *Transforming urban transport*. Oxford University Press.

Guerra, E. (2016). Planning for cars that drive themselves: Metropolitan Planning Organizations, regional transportation plans, and autonomous vehicles. *Journal of Planning Education and Research*, *36*(2), 210–224.

Meijer, A. & Bolívar, M. P. R. (2016). Governing the smart city: A review of the literature on smart urban governance. *International Review of Administrative Sciences*, *82*(2), 392–408.

Millard-Ball, A. (2016). Pedestrians, autonomous vehicles, and cities. *Journal of Planning Education and Research*, 0739456X16675674.

Söderström, O., Paasche, T. & Klauser, F. (2014). Smart cities as corporate storytelling. *City*, *18*(3), 307–320.

# CHAPTER 8

# PLANNING FOR DISRUPTIVE TRANSPORT TECHNOLOGIES: HOW PREPARED ARE AUSTRALIAN TRANSPORT AGENCIES?

John Stone, David Ashmore, Jan Scheurer, Crystal Legacy and Carey Curtis

ABSTRACT

*In Australia, corporations are playing an increasing role in the shaping of urban regions through their ability to mobilize capital to support large infrastructure projects and to usurp institutional planning roles which have traditionally been the responsibility of public-sector agencies. The chapter outlines emerging evidence of changes in the roles of corporations in generating ideas and mobilizing political support for their favoured city-shaping projects, and shows that the private sector is embedded in the processes of government, such as planning, in increasingly complex ways. Through 'market-led' or 'unsolicited' proposal evaluation frameworks, corporations can now bring proposals to political leaders in ways which go outside traditional planning processes and bypass conventional engagement with civil society.*

*In this context, we present data from a recent survey of planners in state and national land-use and transport agencies. The survey, conducted through semi-structured interviews, gathered information about the expectations of these organizations in relation to the nature and timing of the deployment of new AV technologies; about the potential implications for achieving environmental and social planning objectives; and about the collective infrastructure investments that AV technologies may require. This work is being used to shape a new research agenda to explore the planning and regulatory frameworks that are needed to ensure that the AV technologies can be deployed in ways that maximize the public good.*

**Keywords:** Autonomous vehicles; transport planning; Australia; governance; regulation

## INTRODUCTION

Many of the scenarios for AV deployment frame impacts on urban living in simple dichotomies between dystopian and utopian futures. In this context, Isaac (2016) emphasizes a critical role for governance (as well as regulation and policy) in shaping a future involving AVs. As cities continue to grow, citizens and governments are seeking ways to mitigate the effects of climate change, environmental system collapse and the social consequences of rising inequality. The public purpose of urban planning, as Gleeson and Beza (2014) argue, is to facilitate the design and delivery of collective solutions to these and other urban problems. However, urban research currently fails to offer a coherent framework which brings the potentials of AV technology into dialogue with this conception of urban planning.

In Australia, after three decades under a dominant neoliberal paradigm, we can see a closing of political and institutional minds to the possibility of collective effort, and therefore an existential threat to the practice of urban planning. The private sector has become deeply embedded in many aspects of urban transport and land-use planning from the funding of major urban toll roads and the operation, through franchises, of complex suburban rail operations to the management of land-title records and the privatization of building inspections. As 'privatized urban monopolies ... control ever larger parts of Australia's metropolitan estates' (Gleeson, 2017, p. 183) the capacity for state agencies to direct the shape of urban development is further hollowed out (Streeck, 2016, p. 72) and corporate ambition is increasingly fused with state power and resources in what we can call 'corporatized governance' (Paul, 2016, Ch. 2). This is the context in which planners in Australian transport and land-use agencies are positioning themselves to meet the challenges of a possible transformation of urban transport driven through technological innovation generated by the private sector.

Even without such constraints, planning in the face of complex change is a difficult task. Guerra (2016), whose interviews with planners in US metropolitan planning agencies provided the catalyst for our own project, lays down a challenge to planners by reminding them of failures to respond effectively to past technological innovations from the automobile to the postal service:

> *Planners may yet again fail to influence the relationship between cities and a new transportation technology by either misunderstanding driverless cars or seeing them as a solution for contemporary planning problems, such as road congestion or climate change. (p. 211)*

Acknowledging this challenge, we support the collective project of urban planning, if for no better reason than that we believe its laissez-faire alternative offers little hope of achieving our environmental or social aspirations for city survival (see also Chapter 2 in this volume). So, to shape our future research and engagement in the future of Australian cities, we interviewed a small sample of senior staff in Australian transport and land-use planning agencies to better understand their attitudes to the planning and regulation of emerging AV technologies. This chapter reports on the lessons learned in these interviews. We begin by giving some background to progress towards AV deployment in Australia and the environment in which urban planners currently operate as they contemplate the task of managing the driverless city of the future.

## AV DEPLOYMENT AND URBAN PLANNING IN AUSTRALIA: THE CURRENT STATE OF PLAY

The closure of large General Motors and Toyota plants in late 2017 marked the end of local car manufacturing in Australia (ABC, 2017a). So, more than ever, the design of the local car fleet will be decided offshore. However, governments of most States within the Australian Commonwealth are actively competing to attract businesses engaged in the development of AV systems, and trials of various types of vehicles are in progress in several cities (ABC, 2017b; Government of Western Australia, 2016). Local and international telecommunications companies are collaborating with universities in trials of out-of-vehicle communications and guidance systems (Premier of Victoria, 2017) and national agencies are working to:

> *... put end-to-end regulation in place by 2020 to support the safe, commercial deployment and operation of automated vehicles at all levels of automation. (National Transport Commission, 2017)*

Australian drivers – the large majority of whom live in the suburbs of the major cities where a car is a virtual necessity for access to urban life – are becoming aware of AV technology through popular media. And, while only a minority now think that they would purchase an AV, half of the urban population apparently see AVs as a viable alternative to public-transport, if the price was right (ITLS, 2017). These results suggest that international companies wishing to sell AV technologies in Australian cities will eventually need to find ways to persuade initially reluctant drivers to give up control, but that the first fruitful market for these companies might be found in competition with existing public transport services.

The sphere of influence for planners to respond to changes such as AVs has shrunk dramatically in recent decades. We can see this in the example of Victoria, where, as neo-liberalism took hold in the 1990s, the public service was hollowed out and the operation of large metropolitan train and tram systems was franchised. Thus, the capacity for strategic transport planning was restricted, and even the head of the state planning agency argued that planners were relatively powerless:

> *... the importance and – even more, the possibility – of the application of conscious choice to city formation is exaggerated. (Paterson, 2000, pp. 377–386)*

And, today, planners in Victoria work in an environment in which regulation is viewed as 'red tape': an imposition on business rather than a protection for the public (DTF, 2017a). This mindset has permitted the emergence of new processes that undermine traditional approaches to strategic urban planning. It is now possible to fast-track private-sector proposals that have political support (DTF, 2017b). These processes are being used at a city-shaping scale. For example, a current proposal, now in an advanced stage, will permit construction of a AUD 5.5 billion inner-urban toll-road in Melbourne that had not been foreshadowed in any metropolitan strategic plan. If it is completed under current terms, which include mobilization of over AUD 3 billion in private capital, this project would entrench the monopoly position of the highly profitable toll operator. This degree of influence over the use of key parts of the urban road network, combined with a long-term ambition to use its existing tolling technology to roll out electronic road-pricing across the urban road network (Millar & Schneiders, 2016), puts this company in a powerful position to form strategic partnerships with the international developers of AV technologies. These examples illustrate the restricted domain in which Australian planners operate with respect to the private sector and therefore the inherent difficulties that they may face in protecting the public good through the delivery of collective solutions to problems raised in the deployment of AV technologies.

## PREPARED FOR WHAT? THE CONTEXT FOR INTERVIEWS WITH AUSTRALIAN PLANNERS

International consideration of the objectives for governance of the deployment of AVs (ITF, 2017a) points to the need for planners and regulators to evaluate any proposed commercial or public project not only on questions of safety and liability, but also against objectives including:

- reduction of traffic (to foster active travel and mass transit and to support compact urban form);
- reduction of global and local pollutants;
- improvement in accessibility to tackle social inequality.

To avoid the risk of 'greenwash', it is also necessary that any claims on these questions be supported by publicly verifiable data (see, for example, discussions within the International Transport Forum (2017b)).

A further issue for planning and regulation of AVs is the desired relationship between the new technologies and existing transit networks. Some transport planners argue that AVs will be more valuable if they are integrated with traditional public transport services, and with cycling and walking, because this allows for a 'seamless' experience for users and greater efficiencies for the use of each mode (Lindsay, 2016; UITP, 2017). If AVs are to be deployed in ways that provide lower cost structures or greater operational flexibility to replace or rationalize low-patronage traditional services, then it seems logical that transit planning agencies will need a significant degree of control of the information platforms and payment channels needed to provide new multimodal mobility. This is a task that innovative agencies internationally are already well-placed to perform, building on decades of growth in patronage and increasing political and public support. Transit-planning agencies in German-speaking Europe, in particular, have practical experience of fare and service coordination between public and private transit operators and, more recently, with other mobility services such as car and bike sharing (Goodall et al., 2017; VDV, 2015), but this does not appear to be the case in Australia, where urban transit agencies are only recently emerging from decades of declining patronage and fragile political support (Mees & Groenhart, 2012).

To explore these issues, a series of semi-structured interviews were conducted in March and April 2017 with transport and land-use planners and policymakers. The interviews were framed around the question: how prepared are Australian transport planning agencies for disruptive transport technologies? The question of preparedness in relation to AV technology is obviously made more complex by the uncertainty over the timing and capabilities of any mass market entrant in the field. Guerra (2016), after surveying US metropolitan planning organizations on their approaches to the introduction of AVs, found that:

> *... none of the Metropolitan Planning Organizations most likely to be planning for self-driving cars have incorporated them into their most recent Regional Transport Plans. (p. 213)*

He concluded that even though planners were closely following the evolution of the new technologies, the absence of AVs in their plans was principally due to uncertainty about the impacts of AVs on road capacity, traffic safety, land use and travel behaviour. Our research aimed to explore the extent to which uncertainty was also influencing the responses of Australian planners to AVs.

Participants for the semi-structured interviews were recruited from state or national transport or planning agencies. Six interviews were conducted, with an agreement to anonymity of agency and city. Key themes, discussed in the next section, were identified through an inductive process by which the most important topics were identified by their frequency and the emphasis given to them by interviewees (Boyatzis, 1998; Braun & Clarke, 2006; Fereday & Muir-Cochrane, 2006).

## FINDINGS AND ANALYSIS

### Uncertainty Hampering Planning

Guerra (2016) found that planning agencies in the United States appeared to be failing to prepare for the deployment of AVs because, as one respondent put it: 'We don't know what ... to do about it. It's like pondering the imponderable' (p. 214).

This sentiment was also reflected in our interviews:

> *The planning cycle and the planning horizons have gone even more crystal ball than they ever were before. It's very hard to plan in such an environment. If you try and foresee what's going to happen and try and regulate ahead of the curve, then you are probably going to get something wrong. But then if you leave it too long, then you'll have outdated regulation. [Interviewee 5]*

Respondents felt that getting regulation wrong was a worse problem than having outdated regulation, which could, in theory, be improved upon. So, it was seen as preferable to 'watch and wait' and to maintain dialogue with all parties in the sector, expand and develop the relevant skill and knowledge base within government agencies, and keep the planning process as open as possible.

While uncertainty in relation to the ultimate form of AV deployment and its impact is unavoidable, acknowledgement of uncertainty by planners is not necessarily evidence of unpreparedness. Preparedness might be better gauged by the extent to which planners are working to devise and articulate goals for AVs and to formulate means by which to measure their impact. There was, however, little indication of any push towards an integrated and purposive agenda.

## Government: Enabler or Prescriber?

The degree to which the purpose of government is to prescribe outcomes or enable the market was a recurrent theme, with the prevailing sentiment being that bureaucrats were neither technological nor procurement specialists. Again, the tactic appeared to be 'watch and wait'. Participants consistently pondered the existential choice between laissez-faire and directing the market towards some holistic solution.

> *... there would need to be a certain level of regulation put in place by government pertaining to safety, but otherwise it would be possible for government to step away and allow self-regulation to take place. [Interviewee 6]*

Given the prevailing climate of political and institutional support for the private sector over the public, it is not surprising that interviewees generally accepted that the market be allowed to take things forward with the state taking a light regulatory approach. The exception being safety: there are high hopes that AVs will not only reduce traffic accident rates and severity, but their enormous generation of data will also offer new opportunities to address remaining road safety challenges through regulatory intervention. However, the 'light touch' on questions other than safety is likely to have a significant impact on the evolution of the state planning function. Interviewees, in general, recognized the difference between waiting to see how a sector evolves before enacting an appropriate regulatory framework and allowing matters to be taken forward by the market in a 'hands-off' manner. Most tended to support the dominant ideological view that the government's role should be as an enabler. The implications of this standpoint are discussed in Chapter 1 of this volume and explored through scenarios in Chapter 9.

## Government as a Facilitator of Integrated Supply

Despite the general acceptance of market dominance described in the previous section, some interviewees felt that there was a need to ensure that new modes were integrated into a holistic offering. These respondents suggested that this could be done through a ticketing portal or 'Mobility as Service' application, with coordinated (and possibly subsidized) supply for disadvantaged communities or regional and remote areas (see Chapter 3 for further discussion on the hopes and realities of MaaS). These interviewees felt there was an imperative for the state to prevent supply-side fragmentation, and to manage social dis-benefits:

> *If (AVs) become part of an overarching public transport network, then what is the regulation guiding that? The policy certainly needs to consider how much is left to an organic market-driven response compared to how much is tied in to a centrally driven public transport service. And what obligations are on the transit authority to have a level of control, to say contract someone to provide a ... subsidised service, what level of regulation is needed for that? [Interviewee 5]*

> *There is a growing focus on engagement ... but ... where the commercial motivators of industry don't always sit well ... I think that government has a role to even the balance ... to ensure that the right outcomes are provided for those less profit-driven areas. [Interviewee 6]*

Given current practices, operationalization of this need to provide a social safety net would most likely be achieved through complex contractual relationships with AV suppliers. As the interviewees themselves agreed, the public sector faces a shortfall in the skills required to do this effectively.

The complexity of this question was illustrated by one respondent who said:

> *... an on-demand model for buses rather than regular services ... would give government an opportunity to re-examine the relationship they have with public transport. [Interviewee 4]*

This was not meant to imply that governments should retreat from providing 'marginal' bus services, but rather that the new economies of AVs will require new responses from governments. This was the closest respondents came to acknowledging the need for AVs to operate in ways that strengthen transit systems as an alternative to the private car, as described earlier.

## Public Sector Knowledge Gaps

Interviewees stressed that in such a specialized area – far more complex than traditional transit – the skills needed to critically appraise the benefits of the new technologies, and how to procure them as part of an integrated solution, are underdeveloped:

> *The capacity of the public sector to understand the drivers that motivate private sector behaviour is going to be really important ... A lot of the expertise is going to be held by the private sector, we actually need to build our capacity to be a really informed consumer of services and advice (and) when the private sector holds expertise that the public sector doesn't, it can be very challenging to drive good ... outcomes. [Interviewee 1]*

Given the ideological climate in which Australian planners operate, it is not surprising that some interviewees felt that the public sector generally does not recognize its own power:

> *I think that the (private sector is) very aware that they need the active cooperation of governments ... probably more than we ... realise that they need this.... [Interviewee 2]*

This comment acknowledges the reality that, despite the rhetoric, businesses do not want the removal of all government regulation, but rather want a predictable environment in which their service can be operated with consistent returns. It also speaks to the impacts of 'corporatized governance' through which once-clear demarcations of public and private-sector roles in policy development, planning and system operations and maintenance are being re-organized in response to technological disruptions.

## Fragmentation of Effort

Interviewees recognized several lines of fragmentation: between politicians and planners; between state and national governments; and within the emerging AV industry itself:

> *... all policy development requires an authorising environment ... As a bureaucracy, we're limited by the ambitions of our political masters and their willingness to (explore) innovative areas. [Interviewee 1]*

> *Clearly, there can be competition for investment and jobs and the like, that's what governments do. But when it comes to the actual technology, the way it's being deployed, there's got to be national consistency. [Interviewee 3]*

> *There are multiple portals in place and perhaps there could be better governance to establish the links between those portals ... and a public description of what each of them is focussed on, but there's a bit of competitive tension at the moment which isn't a bad thing. [Interviewee 6]*

Respondents' views on the issues requiring a regulatory framework covered a broad range, including an imperative to adhere to emerging global vehicle and technology standards, given the absence of locally manufactured vehicles and the relative isolation from global supply chains. The integration agenda, identified as critical in the literature briefly surveyed in our earlier discussion of the international context for the interviews, was not explicitly recognized as such by most of the respondents.

## CONCLUSIONS

The uncertainty about the form of the new technologies that Guerra (2016) found to be causing paralysis of action among metropolitan regional planners in the United States was also found in Australia, but the uncertainty goes further to encompass the almost existential crisis of legitimacy and purpose that Australian planners face after decades of neo-liberal doctrines in planning and governance more generally. So, planning institutions as a whole are more likely to be avoiding some of the more difficult challenges. This is unfortunate as there is only a short time available in which planners will be able to formulate possible policy and regulatory responses before the emergence of commercial proposals for permission to deploy new vehicle technologies and requests for such things as state provision of communications support or preferential access to road space.

The evidence from our interviews suggests that some Australian planners are grappling with these questions, but, outside the questions of safety being pursued by the National Transport Commission, these individuals are only just beginning to articulate the formal and informal processes by which policy and regulation to achieve these outcomes can be created. Further research and engagement is needed to deepen understandings of planners' attitudes and to participate in the development of mechanisms for management of the deployment of AVs. This can be used to open a wider public debate, with all sectors of the AV industry and with civil society, to build consensus around new requirements for planning and regulation.

## ACKNOWLEDGEMENTS

The authors wish to recognize the generous contribution of the anonymous interviewees. This chapter is adapted from a paper presented at the *State of Australian Cities* Conference in Adelaide in November 2017.

## REFERENCES

[ABC] Australian Broadcasting Corporation. (2017a). Retrieved from http://www.abc.net.au/news/2017-10-02/toyota-and-holden-factories-to-shut-down/9008312. Accessed on 25 October 2017.

[ABC] Australian Broadcasting Corporation. (2017b). Driverless vehicle technology to be tested at Adelaide Airport, Flinders University, 21 March 2017. Retrieved from http://www.abc.net.au/news/2017-03-21/driverless-car-trial-atadelaide-airport,-flinders-university/8373006. Accessed on 22 September 2017.

Boyatzis, R. E. (1998). *Transforming qualitative information: Thematic analysis and code development*. Thousand Oaks, CA: Sage Publications.

Braun, V. & Clarke, V. (2006). Using thematic analysis in psychology. *Qualitative Research in Psychology*, *3*(2), 77–101. doi:10.1191/1478088706qp063oa

DTF. (2017a). Department of Treasury and Finance. Retrieved from http://www.dtf.vic.gov.au/About/Structure-and-functions/The-Red-Tape-Commissioner/Red-Tape-Rubbish-Bin. Accessed on 3 June 2017.

DTF. (2017b). Retrieved from http://www.dtf.vic.gov.au/Infrastructure-Delivery/Market-led-proposals. Accessed on 3 June 2017.

Fereday, J. & Muir-Cochrane, E. (2006). Demonstrating rigor using thematic analysis: A hybrid approach of inductive and deductive coding and theme development. *International Journal of Qualitative Methods*, *5*(1), 80–92. doi:10.1177/160940690600500107

Gleeson, B. (2017). The metropolitan condition. In S. Hamnett & R. Freestone (Eds.), *Planning metropolitan Australia*. Abingdon: Routledge.

Gleeson, B. & Beza, B. B. (Eds.). (2014). *The public city: Essays in honour of Paul Mees*. Melbourne: Melbourne University Press.

Goodall, W., Fishman, T. D., Bornstein, J. & Bonthron, B. (2017). *The rise of mobility as a service: Reshaping how urbanites get around*. Deloitte University Press. Retrieved from https://dupress.deloitte.com/dup-us-en/deloitte-review/issue-20/smart-transportation-technology-mobility-as-a-service.html

Government of Western Australia. (2016). Trial of autonomous vehicles. Media Statements, 9 February 2016. Retrieved from https://www.mediastatements.wa.gov.au/Pages/Barnett/2016/02/Trial-of-autonomousvehicle.aspx. Accessed on 22 September 2017.

Guerra, E. (2016). Planning for cars that drive themselves. *Journal of Planning Education and Research*, *36*(2), 210–224. doi:10.1177/0739456X15613591

Isaac. (2016). *Driving towards driverless: A guide for government agencies*. New York, NY: Parsons Brinkerhoff, WSP.

[ITF] International Transport Forum. (2017a). Retrieved from https://2017.itf-oecd.org/outputs-resources

[ITF] International Transport Forum. (2017b). Retrieved from http://2017.itf-oecd.org/new-business-models-authorities

[ITLS] Institute for Transport and logistics Studies. (2017). Retrieved from http://sydney.edu.au/business/__data/assets/pdf_file/0005/330188/TOPS-2017-Q3_Final_noITLSlogo.pdf, Sydney University.

Lindsay, G. (2016). Now arriving: A connected mobility roadmap for public transport. New Cities Foundation. Retrieved from http://www.newcitiesfoundation.org/wp-content/uploads/2016/10/PDF-Now-Arriving-A-Connected-Mobility-Roadmap-For-Public-Transport-Greg-Lindsay.pdf

Mees, P., & Groenhart, L. (2012). *Transport policy at the crossroads: Travel to work in Australian capital cities, 1976-2011*. Melbourne: RMIT.

Millar, R. & Schneiders, B. (2016). Transurban: The making of a monster. *The Age*, 14 May 2016.

National Transport Commission. (2017). *Changing driving laws to support automated vehicles*. Discussion Paper, October 2017, NTC, Melbourne.

Paterson, J. (2000). Choice and necessity in metropolitan planning. *Urban Policy and Research*, *18*(3), 377–386.

Paul, E. (2016). *Australian political economy of violence and non-violence*. London: Palgrave Macmillan UK.

Premier of Victoria. (2017). Retrieved from https://www.premier.vic.gov.au/world-first-hi-tech-transport-research-for-melbourne/. Accessed on 25 October 2017.

Streeck, W. (2016). *How will capitalism end?* New York, NY: Verso.

UITP. (2017). Autonomous vehicles: Potential game changer for urban mobility. Retrieved from http://www.uitp.org/news/autonomous-vehicles-urban-mobility

[VDV] Verband Deutscher Verkehrsunternehmen. (2015). *Zukunftsszenarien autonomer Fahrzeuge. Chancen und Risiken für Verkehrsunternehmen*. Positionspapier, November. Retrieved from https://www.vdv.de/position-autonome-fahrzeuge.pdfx (in German).

# CHAPTER 9

# DOES GOVERNANCE MATTER? AN INTERNATIONAL SCENARIOS EXERCISE

Greg Marsden and Louise Reardon

ABSTRACT

*As with previous transport innovations, the transition to 'smart mobility' will occur in different ways and at different speeds in different places. Innovations such as Uber and trials of autonomous vehicles are already being welcomed in some places but resisted in others or left to the market. While the technologies may have the potential to be deployed globally, how this happens is, in part, down to the institutional settings and approach to governance amongst all of the actors (public and private) involved. Deciding who should act, how, when and at what spatial scale is, we argue, critical in setting the conditions in which new mobility systems can flourish but in a way which promotes the goals of local, state and federal governments and meets the needs of citizens as well as the industries that promote them.*

*This chapter reports on an international scenarios exercise conducted in 2017 across nine countries. Key dimensions of uncertainty were the degree of governmental involvement in steering policy and the degree of social desirability for smart mobility innovation. Reflecting on the period up to 2035, the scenarios considered the implications for smart mobility transitions by asking which innovations are more likely to flourish and which falter. Strong state involvement is reported as a necessary condition for the most integrated and sustainable visions of smart mobility. Other pathways were suggested to favour some innovations over others but typically offer a smaller market and more atomized and less sustainable set of mobility options.*

**Keywords:** Mobility; governance; scenarios; international; transition

## INTRODUCTION

Nations with developed economies have been through a major transition to what Urry (2004) describes as a system of 'automobility' where the car has become intricately interwoven into every aspect of society and culture, demanding new infrastructures and creating 'private' mobility. There exists however today a debate about a new mobility transition which could take society away from the congestion, inducing and environmentally damaging individualized mobility system that exists today towards a cleaner, more efficient and inclusive future (Bakker, Maat & van Wee, 2014; Docherty, Marsden & Anable, in press; Fagnant & Kockelman, 2015). This we refer to as the smart mobility transition. It comprises several elements which are often imagined together in conceptualizing how the transportation system might change. Critical elements of this are a shift to more autonomous and connected vehicles, moving away from fossil-fuelled vehicles to electric or hydrogen-powered vehicles and increasing sharing of vehicles through new mobility services or peer-to-peer sharing facilitated through mobile Internet. New aggregators of mobility services are also emerging, seeking to replace some or all of the individual ownership model with a subscription-based 'Mobility as a Service' approach (MaaS approach) which looks to find the best personalized options for specific journeys, often through greater multi-modality.

While this is all technically very challenging, it has the potential to offer significant benefits to transportation users and policymakers. The

International Transport Forum (2017) estimate that replacing the private car in the Lisbon metropolitan area with a coordinated 8- and 16-person shared taxi service with very high levels of service could improve journey quality, reduce vehicle kilometres by 48% and $CO_2$ emissions by 62%. Automated vehicles are also seen to offer potential benefits to cities, such as vastly reducing parking and creating better urban realm (Hargrave, 2013). However, whether these systems offer benefits in theory is different from whether they will in practice. This is where questions of government, institutional capacity and governance of a new network of actors become critical. Whether the smart mobility transition will deliver a more sustainable and inclusive future will, in part, be defined by how the systems are allowed to develop and deploy.

Looking again at the Lisbon example above, the conditions which appear necessary to achieve system-wide benefits are very difficult to imagine. First, they require one overarching dispatch agent (so one provider who dictates service quality) and no competition on provision. Second, they require almost all people to not use their own vehicles, with the benefits disappearing entirely if cars are used on average three days a week. The system relies on a very strong societal shift from individualized ownership and use to both collective vehicles and shared vehicles. By contrast, the development of services such as Uber and Lyft resulted from a dissatisfaction with the way in which the taxi market was regulated (by the state) and the opportunity to do things differently (Dewey & Rayle, 2016). These new providers are looking to create new markets with better, continuously developing services. Rather than a market of single providers, we are seeing competition between firms for what is seen to be a lucrative market for mobility (Wockatz & Schartau, 2015) and many different actors seeking to redefine their place in the mobility market including car manufacturers and public transport providers. Central to any consideration of the potential benefits of smart mobility must be how the transition is managed. While desirable end states might be imagined, they might not be attainable (see also discussion in Chapter 2).

This chapter focuses on the importance of governance to the transition through a scenarios exercise conducted with 15 participants from nine countries. The exercise explored how different governance approaches would influence the adoption of key smart mobility system elements.

We briefly, given earlier contributions in the book, elaborate our understanding of governance. The scenarios exercise is then presented. Key findings are discussed in terms of the expected outcomes of alternative futures on system adoption. The chapter concludes with a reflection on what this tells us about the critical elements of governance in steering the transition.

## GOVERNANCE AND MOBILITY

We take our understanding of 'governance' from Treib, Bähr, and Falkner (2007, p. 3) who define it as the 'steering and co-ordination of interdependent actors based on institutionalized rule systems'. This understanding contains three relevant dimensions. First, there is an increasingly complicated and mixed network of actors (state and non-state) involved in shaping and delivering policy. Second, there are rules, laws and formal and informal practices that determine the provision and management of services. Third, governance is a 'mode of political steering' as opposed to 'rowing', where the key focus is on the development of policies to steer progress towards a set of policy objectives, rather than the delivery of these policies by the state alone. Indeed, as Rhodes (2007, p. 1244) notes, the task of modern governance can therefore be understood as the expression of a set of 'beliefs, practices, traditions and dilemmas' that are negotiated democratically, and therefore requires a dialogue with the public on what they find to be acceptable conditions, practices and consequences of action.

The government, representing the nation state, is but one actor in this governance mix. However, while the role of the state is evolving, it remains a critical actor through its role as an 'enabler', stimulating others to action, and 'ensurer', obligated to make sure action achieves its defined outcomes (Giddens, 2008). In this latter role, in particular, the government is '...fulfilling roles which other actors normally cannot; ...Accountability for services and performance, and responsibility especially if things go wrong' (Giddens, 2008, p. 9).

Governments currently can and do intervene in the existing mobility system in numerous ways (Gudmundsson et al., 2015). For example, they set transportation policy goals and priorities to ensure that the mobility system supports other key policy goals relating to economic and social progress. They also set standards on different levels of transportation service, communicating these with the public in order to justify taxation spending. Governments also intervene to address market failures. For example, they act as providers or procurers of services which are not profitable for private providers in the market environment and they regulate to limit market power of monopolistic or oligopolistic providers. They also set basic standards of operation and rules of movement between transport modes and infrastructures through tools of the legal system, in order to ensure a basic level of interoperability. Governments also intervene by way of investment: funding the provision and upkeep of infrastructure, and acting as guarantor of public services when private provision fails.

The extent and variety of ways in which governments intervene in the existing mobility system through steering governance suggests that they are not going to be any less influential in shaping the future smart mobility transition. Indeed, one of the distinctive factors in creating sustainable transitions is that they require strong input from the state and civil society because it is unlikely they 'will be able to replace existing systems without changes in economic frame conditions (e.g., taxes, subsidies, regulatory frameworks) … (which) will require changes in policies' (Geels, 2011, p. 25). However, the role of the state will likely become tested and stretched as more new actors enter the smart mobility space, bring with them new resources and data which affect the existing balance of power within governance networks, and which challenge existing ways of working.

It is also important to acknowledge (see also Chapter 2) that the smart mobility transition is happening alongside parallel transitions in society in understanding the role of the state. A post-recessionary shift to the political right is accelerating programmes to reduce the size and role of the state in decision-making and social support and to provide new delineations about both what the state should intervene in and why it should do so. This is particularly notable in transport, with the increased focus on mobility and mobility services as an industrial and growth policy as much as a transport policy (see also Chapters 5 and 8). The position of the state and the scope for action should, therefore, be considered from both normative and pragmatic perspectives.

## THE SCENARIOS EXERCISE

Zegras, Sussman and Concklin (2004, p. 2) summarized scenario planning as 'a tool designed to help an organization judge how effective decisions made today will be in the uncertain future. Scenario planning is not a replacement for traditional planning techniques such as forecasting; instead it aims to help organizations better prepare for the unexpected. In short, scenario planning helps to make robust strategic choices'. We suggest that the current environment does indeed suggest a set of uncertain futures. Even with one single technology such as automation, there is little agreement as to the degree to which full automation will be available or the pace at which it will happen.

Here we adopt an explorative strategic scenario approach which have been shown to be useful 'in cases when the user may have fairly good knowledge regarding how the system works at present, but is interested in exploring the consequences of alternative developments' (Börjeson et al., 2006, p. 727). The approach to developing scenarios involved identifying two important, uncertain and independent factors. These are plotted on two perpendicular axes to produce four quadrants, each representing alternative futures of the world which could unfold (Figure 1).

Informed by our review of governance literature we selected:

1. 'Hands-on' versus 'Hands-off' government. 'Hands-on' government refers to a set of governmental institutions that have the capacity and will to exercise the range of tools of government (e.g., taxation, regulation, co-ordination and investment) while 'hands-off' government refers to a lack of capacity and/or will to exercise these tools.

2. Degree of social acceptance of innovation. The positive side of the axis could be interpreted to include widespread acceptance of robots, automation, drones, sharing, data sharing, electric vehicles or other new forms of propulsion technologies. The negative side of the axis would, by contrast, see cautious attitudes to data sharing or sharing more generally, resistance to automation of activities amongst significant parts of the population, but not all.

Note, in line with all scenario exercises, these are supposed to depict plausible futures and not necessarily desirable futures. The purpose of picking governmental approaches and societal acceptance as the two axes

**Figure 1.** The Four Scenarios.

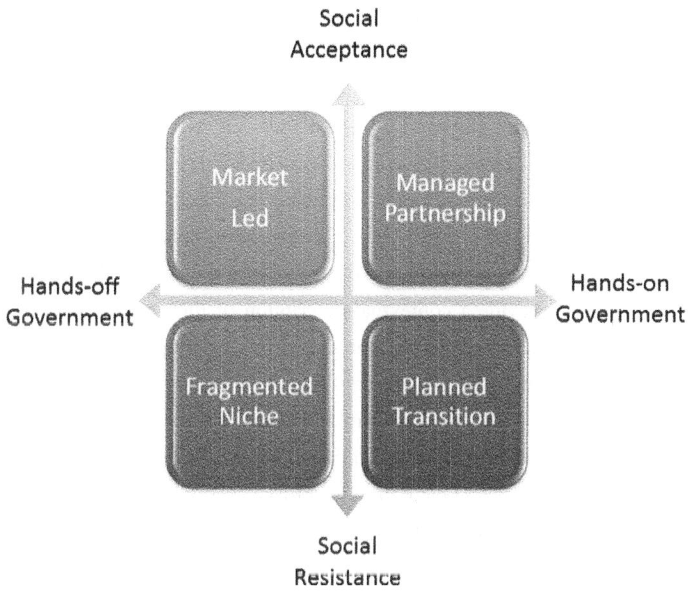


was to explore the extent to which participants in the scenarios exercise felt that the actions of the government, in different societal settings, might open or close pathways for certain types of innovations or combinations of innovations. This captures both the importance of different approaches to governing and, as noted by Rhodes, the importance of dialogue with the public and democratic accountability.

A short description of each of the scenarios was developed, which gave some contextual information, such as the importance of climate change legislation (stronger in hands-on government) and the tax base and capacity of the government to respond to changing circumstances (stronger in hands-on). The descriptions also drew attention to more general contextual changes such as an aging population and the need to think about the development pathway of innovations beyond the specific boundaries that the respondent was familiar with. Some of the innovations might be allowed to flourish in some parts of the world for example and this may create external pressures to adopt them.

The scenarios were circulated amongst the World Conference on Transport Research Society Special Interest Group on Governance and Decision-Making and all registered attendees at the International Transport Forum's pre-summit research day on Smart Mobility Governance. Fifteen responses were received from nine different countries (Australia, the United Kingdom (England and Scotland), the Netherlands, Austria, Poland, the United States, Sweden and Norway). Where multiple responses were received, an aggregation of responses was used to determine the position of the country rather than allowing one country with multiple responses to have undue influence on the overall narratives. Participants were asked to provide a contextual description of their country's governance arrangements and the current approach to the uptake of different aspects of smart mobility. They were then asked to describe whether they felt different innovations (automated driving, electric vehicles, MaaS and integration between systems), in particular, would flourish or falter in the different scenarios. Qualitative narratives were requested for each, which included broader considerations about how the scenario might unfold.

The sample has strengths in being international in composition but it remains limited. A more multi-level understanding of the perspectives of the various state actors at different spatial scales and in different locations within a country would be useful to explore as would greater attention to different types of private-sector actors (incumbent, new provider, manufacturer, system integrator, service aggregator).

## RESULTS OF THE SCENARIOS EXERCISE

Figure 2 provides an overall qualitative reflection of the findings regarding which innovations were deemed likely to flourish or falter. Four symbols meant unanimity across the countries about flourishing of an innovation and half a symbol widespread agreement that it would falter. We present here some overarching points from the scenario aggregation before reviewing each scenario. However, we stress the importance of start point context (such as the degree of car dependence, the extent to which government takes a strong, market-led position to governance approaches and the extent to which the electricity grid is coal based) which can make a

**Figure 2.** Scenario Outcomes – Innovations That Flourish and Falter.

*Key*: Battery car = electric; Mobile phone = MaaS;
Wi-Fi car = automated vehicles; Jigsaw = integrated mobility.

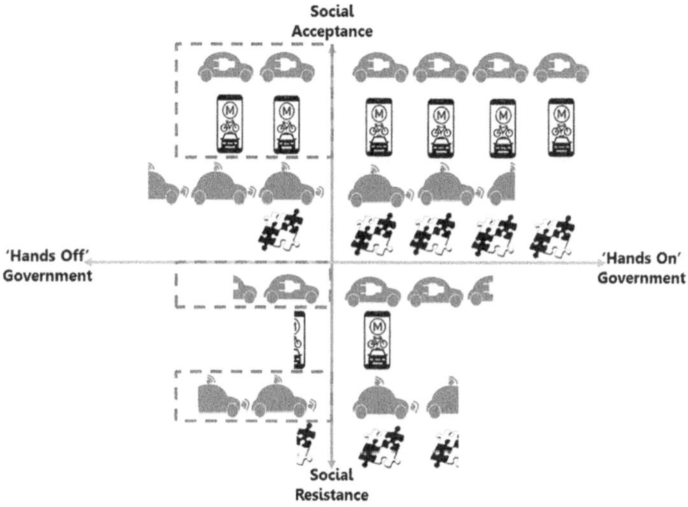


substantial difference about which innovations are perceived as likely to succeed.

The combination of hands-on government and a social desire for an integrated system in the Managed Partnership scenario (top right of Figure 2) allows all of the systems to flourish. There was debate about the extent to which an early adoption of integrated smart mobility with effective MaaS products might diminish the market for automated solutions as reliance on the car was generally reduced. A strong state would ensure pricing co-ordination, integration and reform to provide a robust fiscal base for policy in the light of the shift to electrification. Anticipatory regulation and active piloting of new systems would be used to accelerate deployment. A transition in how what 'public transport' was and how it was provided would be managed. Corridor services would remain important, but other services would be more blurred. Within this scenario, the state had tools to assure socially necessary provision for citizens and areas where it may otherwise not be commercially viable.

The planned transition (bottom right of Figure 2) led to a managed introduction of a more individualized approach to travel with some successes in shared mobility for some groups and places where there was a will. But in general this scenario achieved little in the way of integration – partly because there was a lack of social desire for it, and government, therefore, did only what it needed to do. Mobility as a Service would be more about access to packages of services so that multiple modes can be used but more across the week or month than for integrated multi-modal journeys. The division between private and public transport would remain strong. Car-based solutions would remain dominant for rural and poorly served areas.

Looking at the 'hands-off' scenarios, the lack of willingness or capacity to intervene in both of the hands-off scenarios leads to an individualized approach to mobility coupled with some AV and EV adoption. In the Market-led scenario (top left of Figure 2) there was strong proprietary system development and competition between cities for systems and development opportunities which could lead to lowest common denominator regulation. There was a view that there would be more limited integration across products and a decline in public transport due to system competition. The fully intermodal world of the managed partnership was not attainable as regulatory barriers were not cleared out of the way. Some areas would be poorly served by public transport and new mobility services (rural, sub and peri-urban) and some user groups might be excluded or disadvantaged. The hands-off approach would fail to tackle pricing reform which, coupled with the limited co-ordination between systems, would result in rising congestion despite uptake of AVs.

There were very pessimistic views on integration and mobility service opportunities in the Fragmented Niche scenario (bottom left of Figure 2). Some respondents suggested that this scenario might most accurately reflect 'business as usual'. A lack of public support and government commitment would stifle innovation. Developments would happen in niches where there was sufficient commercial value. There would be a rise in individualized AVs and EVs but more limited to higher earners. There would be a decline in traditional public transport services under reduced funding and as parts of the market are chiselled away by niche services. Some respondents felt that the failings in the mobility system may generate incentives to promote more localized lifestyles, walking and cycling.

The boxes with dashed lines in Figure 2 correspond to the areas where there was considerable disagreement amongst the participants. So, the extent to which electric vehicles would accelerate where limited state intervention was likely in the Market-led (top left) scenario was questioned, as was the extent to which MaaS was possible to create outside a stronger government framework. There was disagreement on whether AVs were likely in more than a few specific circumstances in the Fragmented Niche scenario (bottom left). There too, electric vehicles were seen by some to be more limited and by others to be a natural option for people seeking to escape failed transport integration. The aggregated messages from a scenario exercise clearly hide some important nuances. For example, in the United Kingdom a commitment to all new cars being fully electric by 2040 has been made yet it seems less clear that the competition issues facing a fully integrated MaaS offering will be tackled.

## CONCLUSIONS

This chapter set out to explore the contention that the smart mobility transition was, in fact, many transitions. The nature and speed of any transition will be determined, in part, by the context in which it happens. Through a scenarios exercise, this chapter has established that two key dimensions of that context are the actions of government and the degree of societal acceptance and impetus for different parts of the transition.

It is clear that the system and networks of governance are broadening, and the challenges of steering policy through a more complex implementation environment are becoming greater in the field of transport. Nonetheless, the approach of government still appears likely to matter very significantly to how system adoption will unfold, with important implications for industry and society. More hands-on government will be necessary to stimulate early adoption of some innovations, to tackle reforms to pricing and to manage co-ordination between new and 'old' mobility service providers and to manage this in ways which protect the interests of all citizens. More individualistic and niche system futures also remain plausible. These seem likely to reduce the full scope for the markets to develop, being confined to more affluent niches and areas where commercial economies of scale work. In many senses these are old governance problems but, in the face of declining revenue from fuel taxes and tightened public spending, the implications could be more severe for those reliant on public transport, living in places where the economies of scale or other costs of providing basic universal delivery do not make sense.

Finally, our findings point to an urgent need for effective comparative analysis. There already exists a wide range of styles of governance in different institutional settings. Tracing what governance strategies work with which innovations over space and time is a critical step in refining our understanding of the role of governance in shaping this transition.

## ACKNOWLEDGEMENTS

We would like to thank the International Transport Forum for the funding they gave to the initial research day from which this collaboration emerged. We would also like to thank our co-authors in the book for their time and intellectual input to making this book what it is.

## REFERENCES

Bakker, S., Maat, K. & van Wee, B. (2014). Stakeholders' interests, expectations, and strategies regarding the development and implementation of electric vehicles: The case of the Netherlands. *Transportation Research Part A: Policy and Practice*, 66, 52–64. doi:10.1016/j.tra.2014.04.018

Börjeson, L., Höjer, M., Dreborg, K. -H., Ekvall, T. & Finnveden, G. (2006). Scenario types and techniques: Towards a user's guide. *Futures*, *38*, 723–739. doi:10.1016/j.futures.2005.12.002

Dewey, O. F. & Rayle, L. (2016). How ridesourcing went from 'rogue' to mainstream in San Francisco. Retrieved from http://research.gsd.harvard.edu/tut/files/2016/06/San-Francisco-Case-2016.pdf

Docherty, I., Marsden, G. & Anable, J. (in press). The governance of smart mobility. *Transportation Research Part A; Policy and Practice.* doi: https://doi.org/10.1016/j.tra.2017.09.012

Fagnant, D. & Kockelman, K. (2015). Preparing a nation for autonomous vehicles: Opportunities, barriers and policy recommendations. *Transportation Research Part A: Policy and Practice*, 77, 167–181.

Geels, F. (2011). The multi-level perspective on sustainability transitions: Responses to seven criticisms. *Environmental Innovation and Societal Transitions*, *1*(1), 24–40. doi:10.1016/j.eist.2011.02.002

Giddens, A. (2008). *The politics of climate change*. London: Policy Network.

Gudmundsson, H., Hall, R. P., Marsden, G. & Zietsman, J. (2015). Sustainable transportation indicators, frameworks, and performance management. Retrieved from https://link.springer.com/book/10.1007%2F978-3-662-46924-8

Hargrave, J. (2013). Driverless cars to end parking problems. Retrieved from http://thoughts.arup.com/post/details/273/driverless-cars-to-end-parking-problems

International Transport Forum. (2017). Transition to shared mobility: How large cities can deliver inclusive transport services. Retrieved from https://www.itf-oecd.org/sites/default/files/docs/transition-shared-mobility.pdf

Rhodes, R. A. W. (2007). Understanding governance: 10 years on. *Organization Studies*, *28*(8), 1243–1264. doi:10.1177/0170840607076586

Treib, O., Bähr, H. & Falkner, G. (2007). Modes of governance: Towards a conceptual clarification. *Journal of European Public Policy*, *14*(1) 1–20. doi:10.1080/13501760601071406

Urry, J. (2004). The 'system' of automobility, theory. *Culture and Society*, *21*, 25–39. doi:10.1177/0263276404046059

Wockatz, P. & Schartau, P. (2015). IM Traveller Needs and UK Capability Study: Supporting the realisation of intelligent mobility in the UK. Retrieved from https://ts.catapult.org.uk/current-projects/traveller-needs-uk-capability-study/

Zegras, C., Sussman, J. & Conklin, C. (2004). Scenario planning for strategic regional transportation planning. *Journal of Urban Planning and Development*, *130*(1), 2–13. doi:10.1061/(ASCE)0733-9488(2004)130:1(2)

# SECTION FOUR

# CONCLUSION

# CHAPTER 10

# CONCLUSION: A WINDOW OF OPPORTUNITY

Louise Reardon and Greg Marsden

ABSTRACT

*This chapter considers the book as a whole to draw out the key findings and link these to broader themes. In it we suggest that as we are in the relatively early stages of smart mobility adoption, there exists a key window of opportunity to think about to what end, and how, the transition can be steered. If this window is not seized, there is a heightened risk that governments will always be responding to, rather than shaping, events.*

*Four key themes emerge from our reflection. First, while the current system is not equitable, there is a need to be cognizant of who the new winners and losers will be in the next transition. It is far from clear that commercial business models always align with notions of social provision. Second, while the early stages of the transition appear to be marked by technological optimism, reinforced by government industrial strategies, it seems clear that without politics with a capital P, the opportunity to engage the public and educate decision-makers will be missed. This leads to our third point: that a lack of pro-active state engagement in the early stages of these innovations will undermine the institutional capacity to engage later on in the transition. Without governmental steering, the transition risks representing what we are given rather than what we want or need it to be. Finally, we highlight the importance of context to how the transition will unfold, with some countries or cities already at arm's length from transport provision while others are more hands-on.*

*The chapter concludes with our reflections on what actions governments could take now to prepare better for the transition and on what the volume of work says for future research needs. While greatly enthused and rewarded by the debates we have had in assembling this book, we see these as the starting point for future agendas and very much open to contestation. We, therefore, hope this book acts to advance the study of the governance of smart mobility and to elevate its status relative to the significant body of work underway on the technologies themselves.*

**Keywords:** Politics; transition; steering; equity; capacity

## A 'WINDOW OF OPPORTUNITY'

This edited collection is published against the backdrop of growing momentum for smart mobility services and products; Uber's brand has entered the lexicon of contemporary city life; high-profile media commentaries on automated vehicle testing are commonplace; and demand for Tesla continues to grow. In framing the book as we have – around governing the smart mobility transition – it therefore makes sense to reflect on where in the *transition* we are, stepping back from the hype surrounding smart mobility, in order to then identify the opportunity to 'govern' such change. Meadowcroft (2009, p. 324) defines transitions 'as processes of structural change in major societal subsystems. They involve a shift in the dominant "rules of the game", a transformation of established technologies and societal practices, movement from one dynamic equilibrium to another – typically stretching over several generations'.

As has been acknowledged across the contributions, to frame smart mobility as one 'transition' is a misnomer – more accurately smart mobility involves transitions, plural; MaaS, electric vehicles, automated vehicles and ride-sharing, for example, have all been highlighted as involving their own complex networks, technologies and societal responses that embody different aspects of 'smart' and indeed 'mobility'. However, while some of these transitions could be described as 'further along' than others – ride-sharing in contrast to automation, for example – what the contributions highlight in their differing ways is that the 'rules of the game' that have pervaded the automobility paradigm of the last fifty years or more have not yet been torn up or even revised to any great extent, but rather are informing the ways in which smart mobility innovations are framed, responded to, steered and, in turn, will evolve. This book therefore posits that, at the moment, there is a 'window of opportunity' to pause and reflect: learning lessons from the governance of automobility and the challenges and changes it has created, and drawing on instances of smart mobility application, to understand the gaps, opportunities and ability to steer transitions into the future. In attempting to open this window if just a little, the collection provides four take-home messages.

## THERE WILL BE WINNERS AND LOSERS

Each of the chapters recognizes, in its own way, that there will be winners and losers both during and 'after' the smart mobility 'transition'. We should not be surprised by this as the current system does not provide an equitable distribution of mobility options or of benefits and disbenefits. Set against this backdrop, smart mobility could resolve some of these long-standing issues. However, it may also perpetuate or exaggerate them. The chapter by Gironés and Vrščaj is the most explicit in examining this issue. Their comparative analysis of two policy interventions highlighted how smart mobility technologies (connected e-bikes) are being utilized to address the externalities of automobility; facilitating behaviour change to create positive societal 'wins', including improving the health of the individual while at the same time reducing congestion (and its associated externalities) for the cities enacting the programme. However, in the application of these interventions the authors highlight how groups of individuals are treated unequally as a means to this end.

The chapters by Docherty, and Marsden and Reardon, highlight the more systemic implications of smart mobility for creating winners and losers. Docherty, for example, concludes that 'Smart providers want to maximize our consumption of mobility to maximize their own returns, and at the same time externalize as many of the costs as possible onto others'. While, in thinking through different governance scenarios for smart mobility in Chapter 9, Marsden and Reardon identify how an upswing in the popularity for automated vehicles, and ride-share services, risk perpetuating an individualized, 'on-demand' culture for transport, which undermines support for the provision of mass (public) transport as demand for such services declines and in turn undermines its financial viability. Such a trend has the potential to reinforce the disadvantages suffered by poorer members of the population (who can't afford to buy or access these services) and those living in rural areas (where the services or infrastructures may not be in place to be taken advantage of). We do not imply that there will be no overlap of commercial viability and social need but suggest that such an overlap should not be taken for granted or its scope exaggerated, given the behaviours of commercial public transport operators to date.

Many of these smart mobility services also require access to different technology platforms (the Internet and smartphones) – potentially inaccessible to lower income members of society – and a level of technological literacy that may be a barrier to older members of society. In relation to MaaS, for example, Pangbourne et al. note that 'its reliance on registration and digitalization...create[s] additional barriers for those who are already experiencing exclusion, adding a loss of mobility to problems caused by, for example, the digital gap or through lack of access to banking'. And also notes that inequalities can be reinforced through business models that offer discounts for smartcard and App users, leaving those using traditional payment methods paying more. That being said, there is also recognition across the chapters that smart can be an enabler for different communities; automated vehicles, for example, may provide a level of personal mobility and freedom that was previously unattainable for those with disabilities, and improve safety for all road users.

The crux here is that the 'winners' and 'losers', be these defined at the individual, group, nation or societal level, are yet to be clearly identified; the distribution of the wins and losses is uncertain; and the winners and losers are likely to change during the transition and within different contexts where transitions occur. The contributions, therefore, highlight the unique role the state has, and must play here, in ensuring that customers and citizens are protected from the negative externalities of the transition. As Docherty argues; 'the key role of the state is therefore to ensure the negative consequences of such increased mobility (as has been experienced during automobility) are not "locked in" and hard to overcome in future'. While the contributions recognize the growing complexity and asymmetries of power that make such a role more challenging for the state, nevertheless it remains one of its key functions, and interrogation of citizens' interests must remain a live issue despite the technological optimism that many of the contributions highlight (Docherty, 2018; Pangbourne, Stead, Mladenovic, & Milakis, 2018; Hopkins & Schwanen, 2018).

## WE NEED POLITICS WITH A CAPITAL 'P'

Dowling emphasizes the importance of the way interventions and technologies are 'categorized' and 'framed' for determining policy action. In this regard, many of the chapters highlight the power of the neo-liberal framing within the transport sector, and how this understanding perpetuates the assumption that facilitating the private market for services is the most appropriate way to deliver mobility for all. The journey towards a smart mobility transition has the potential to compound and reinforce these understandings through the ways in which private companies such as Lyft and Uber 'sell' their interventions and means to new solutions. While such a marketized set-up has implications for the practical mechanisms of governing and coordinating (see sections below), more insidious is the lack of politics, with a capital 'P' that such a framing fosters.

As Hopkins and Schwanen note: 'as far as automated vehicle developments in the United Kingdom are concerned, there appears to be little space for critical interventions, big picture thinking, reflexivity or indeed failure. This can be attributed at least in part to the way in which the smart mobility transition is framed in the context of neo-liberal ideals and dependence on private sector interests'. From a planning perspective, Stone highlights how 'we can see a closing of political and institutional minds to the possibility of collective effort, and therefore an existential threat to the practice of urban planning'. The risk is, therefore, that while learning may be occurring in the technological space – relating to experiments with autonomous vehicles, or MaaS, for example – there is little opportunity for learning more broadly about the social acceptability or implications of these technologies. At the same time as asking 'can we transition to this technology?' we should also be asking 'do we want to transition to this technology?' and, if so, how?

However, as noted above, the way the issue is framed determines its policy response, and at present the framing is strongly influenced by the smart mobility operators themselves who to a certain extent are pushing at an open-door, with national governments keen to include 'solutions' to safety and accessibility which also promise to stimulate economic growth and support climate change mitigation in their industrial strategies. As a consequence, other voices (and frames) remain largely out of the picture and notably citizens are narrowly defined primarily as 'adopters' (see Hopkins and Schwanen). While this may make sense, after all, there has not yet been time for any negatives of 'smart' to materialize, and therefore no obvious coalitions 'against', this does not mean the conversations should not be had. Indeed, in the face of uncertainty, such conversations help us to navigate forward on the best path possible, given the bounded nature of our knowledge of the future and its potential consequences. Indeed, deliberation stops us from determining the solution to the problem before all solutions have been debated, and problems prioritized. For example, automated vehicles may be the solution to the problem of safety, but there may also be less disruptive and simpler solutions out there. Indeed, automation may be the 'best' solution in terms of safety, but reinforce a dependency on the automobile that perpetuates other more severe and long-term problems.

As Meadowcroft (2009, p. 337) notes 'democratic institutions can provide a framework for taking legitimate decisions about what constitutes the public interest, for enforcing rules, and resolving distributional conflicts. Different interests and perspectives can have their say. There are opportunities to redefine identities and interests, and to construct solutions to societal problems'. However, the current positive framings place 'smart mobility' in a largely technocratic box and indeed, as is starkly outlined in Stone's chapter, is rarely discussed and contested in any meaningful way within the 'transport' context. If government (be this federal, state or local) is to help protect the 'losers', ensure positive social outcomes and, in turn, inevitably provide the context for the 'winners' to flourish, it must, therefore, have a democratic basis on which to do this. As Diane E. Davis notes in her contribution: 'we should be asking citizens and their governing authorities to think about the kind of city they wanted to live in, and then adopt or advocate for the mobility systems most likely to lead us there'. And, as she further notes, 'we should not succumb to technological determinism and let our future cities be the aggregated product of a series of random innovations in the IT industry, appealing as they may at first appear'.

## STEER OR BE STEERED

The above points highlight the role of the state in ensuring social outcomes, and the role of democracy and contested political discourse for providing the guidance and vision for what those outcomes should be. However, the networks within which the governance of smart mobility are occurring are growing ever more complex, with growing numbers of actors, more engrained interdependencies and increasingly intertwined modes of delivery. This has the potential, as many of the contributions have noted, to lead to growing power asymmetries, bolstering the power of private mobility providers at the expense of the state. For example, Pangbourne et al. highlight how increased reliability on MaaS increases the potential 'for an entire city to become unable to move, should the MaaS system be compromised, for example, through power failure, ICT failure or a Deliberate Denial of Service cyberattack'. Moreover, in Davis's discussion of Uber in San Francisco she concludes that 'the fiscal gains accruing from this smart mobility innovation remained in private sector hands, and its impacts were meaningful for individual transit consumers but less so local authorities. Overall, there were few positive spillovers for the operation of the entire transport network, either in budgetary or transit coordination terms'. Risks of fragmentation, vulnerability and loss of control are, therefore, very real. However, across the contributions it was noted that it does not have to be this way; the message is steer or be steered.

As discussed above, the question of what government should steer towards falls back on the outcomes of politics with a capital 'P'. However, the point is that once a path has been chosen, the state has the power and tools to navigate on this course should it decide to do so. This power may be embodied through the state acting as a meta-governor – setting the context for governance to occur – rather than through acting directly and unitarily, but nevertheless as Marsden and Reardon highlight in

Chapter 9: 'the approach of government still appears likely to matter very significantly to how system adoption will unfold, with important implications for industry and society'. Indeed, as many of the chapters highlight, the success of numerous smart mobility transitions is predicated on the assumption that the state will embrace a key role in the system: as coordinator (in the case of MaaS for example) or as provider of goods uniquely in their gift (for example, road space in the case of car share). Hopkins and Schwanen's contribution highlights the role of the state in facilitating the drive towards automated vehicles; amending legislation; developing codes of practice; fostering networks and connections between start-ups and incumbents; and supporting technology trials and experimentation. However, it is in the resistance to or rather the steering away from different forms of smart mobility transition that governments may feel uncomfortable. As Stone highlights, such a suggestion may present something of an existential crisis, but not acting is indeed a choice. Moreover, in areas where contestation over implementation has been 'out in the open' as it were, for example, in the protests against Uber in numerous cities, the state has felt comfortable in utilizing its power to 'ban' their operation.

## CAPACITY TO STEER WILL DEPEND ON CONTEXT

Underlying all the above messages, and pervading all the contributions, is the importance of context. Smart transitions are not evolving onto a blank slate; the histories, traditions, cultures and expectations of the individuals, towns, cities, regions and nations in which transitions are occurring will inevitably shape the form they take and the power dynamics that result. Automobility is not the same everywhere. The chapter by Davis highlights the differing norms of Stockholm (with a strong welfare culture with a robust political party system) and San Francisco (an 'outpost' city in a federal system with a commitment to 'decentralized' democracy) as key factors in the governance strategies adopted towards smart mobility innovations in each location. The scenarios presented by Marsden and Reardon highlight how factors such as levels of car dependency within the existing mobility system, reliance on a car manufacturing industry for national GDP and dependence on fossil fuels are also important contributory factors in determining governing responses.

Moreover, the chapters by Davis and Dowling highlight the importance of resisting sweeping generalizations about smart mobilities' impact on state capacity – this is a question of styles of implementation informed by context. In looking at governance approaches, Davis finds that the strategic response to Uber in San Francisco resulted in reduced local-level state capacity in the City, while the strategy of partnership working between Stockholm's city authorities, its Mayor and IBM, led to an opposite outcome. In focusing on the incumbent framings of different contexts, Dowling's chapter highlights how car sharing was facilitated by government in a less contested way than Personal Mobility Devices due to the former's alignment with existing categorizations of transport and mobility.

The case studies in this book, while often prospective in nature, have been set in the governance context of today. While today's conditions have been shown to matter, that context itself is not fixed and it is necessary to consider where it might be heading as a result of changes far beyond just the transport sector. There continues, in developing western economies, to be a significant push towards a smaller state with fewer resources committed to ongoing subsidy of operations. Pressures for greater expenditure on health care and social care are increasing as we live longer, which is impacting transport budgets. There is also an uneven distribution of wealth across generations which may challenge the extent to which taxation can fund expenditure in the way it may have done so in the past. Specifically, in the transport sector, an anticipated shift to electric vehicles will massively reduce the tax received direct from transport with domestic energy taxed at much lower rates. We suggest that under a business as usual scenario, the resources for most state-funded institutions will diminish, as will their technical capacity. Without technical capacity to steer and challenge smart mobility providers, we see little likelihood that politicians will identify a need to act to put the P back into politics in this space, as without a clear policy solution they are unlikely to want to acknowledge a problem.

## SO WHAT?

In the discussion at the International Transport Forum meeting we were asked 'so what?' What should governments do when faced with these dilemmas? As we have said, generic advice runs the risk of missing local context and of expiring quickly. We have, therefore, restricted ourselves to a short set of policy recommendations which we think have some level of universality and longevity:

1. Governments, at all levels, should very clearly set out the policy objectives they are seeking to achieve (such as road safety, congestion reduction or social inclusion). Decisions about whether to welcome or resist different innovations should then be about whether those systems help deliver those outcomes. Examine what forms of consumer protection new smart mobility services will require. For example, what will a socially necessary level of provision look like in the next five years? Should we continue to regulate modes or should it be [Mobility as a] service providers? How is access to, and quality of, a basket of mixed services to be regulated to protect consumers? These questions should be addressed while the systems are niche and while opting in and opting out remain viable options for consumers.
2. Consideration should be given by all parties about how new mobility system trials and reforms can be used to engage users, non-users and decision-makers in a more reflexive way of understanding what society wants from these systems rather than what a small elite think they might want.
3. Re-imagine the role of local, regional and national government in the transport arena in parallel with reforms to mobility systems. Given the long-standing modal silo mentality of transport governance and the challenges to changing that, this is not a small task. However, delivering a more user-oriented system which comprises a new mix of systems may require a more radical re-think of how government structures its approach to setting up and governing through networks. It is also essential that the critical questions of capacity to steer are addressed. How many places will have the necessary skills and know-how to challenge the global actors that could dominate the market? What will be done to support the places that do not or will not have such capacity?

As for a parallel prospective research agenda, our intention has been that this book acts as an initial marker underlining the need for a more consolidated research effort on the governance of smart mobility. We argue that how the smart mobility transition is allowed to happen and is steered is of parallel importance to research into the technological questions which exist about how to make it possible to happen. Each chapter provides new ideas and opportunities which could be mined further and our summing up in this chapter has attempted to frame some bigger themes. The confirmation of the importance of context throughout the case studies suggests that much can be gained by a rigorous approach to cross-comparative analysis of how different governance strategies influence the smart mobility adoption pathway. We hope that in having reached this point in the book, you are convinced of the importance of tackling these research challenges now, rather than waiting to see if and, how so, the state might react to try and reduce any downsides that might accompany the benefits that these new systems will undoubtedly bring.

## REFERENCES

Docherty, I. (2018). New governance challenges in the era of 'Smart' mobility. In G. Marsden & L. Reardon (Eds.), *Governance of the smart mobility transition* (pp. 19–32). London: Emerald Group Publishing Limited. Chapter 2.

Hopkins, D., & Schwanen, T. (2018). Governing the race to automation. In G. Marsden & L. Reardon (Eds.), *Governance of the smart mobility transition* (pp. 65–84). London: Emerald Group Publishing Limited. Chapter 5.

Meadowcroft, J. (2009). What about the politics? Sustainable development, transition management, and long term energy transitions. *Policy Sciences*, *42*, 323–340.

Pangbourne, K., Stead, D., Mladenovic, M., & Milakis, D. (2018). The case of mobility as a service: A critical reflection on challenges for urban transport and mobility governance. In G. Marsden & L. Reardon (Eds.), *Governance of the smart mobility transition* (pp. 33–50). London: Emerald Group Publishing Limited. Chapter 3.

# INDEX

Printed and bound by CPI Group (UK) Ltd, Croydon, CR0 4YY

13/09/2026

14961060-0002